Contents

Preface

Purpose of This Book

This book is about media and society in the Digital Age—an age in which many forms of communication, commerce, and content are digitally based, traveling as bits and bytes on global computer networks. In the media industry, it could involve computer-based research and collection of information, production and dissemination of content through electronic networks, creative presentation using a wide range of software applications and hardware tools, and so forth.

The concept of *media* used in this book is broad. At one time, a study of media would entail looking at books, newspapers, magazines, radio, and television—what we might call "traditional mass media" or "legacy media" today. These are specific technologies traditionally used for the dissemination of news, information, opinion, and entertainment. The textbook model of the traditional mass communication process shows a single communicator, or sender, and many receivers (e.g., an audience, readers, viewers, etc.) receiving content through a particular channel. The emergence of the digital media environment challenges this traditional model as well as many conventional notions of media. Students of media and society in the Digital Age—be they in communication(s), journalism, sociology, political science, or any number of other fields—should thoroughly understand the transformations in their society currently being ushered in by changes and innovations in communication and information technology.

In addition to learning about digital media, students (as emerging scholars and researchers) should be critical of developments in the Digital Age from a position of knowledge and not ignorance or irrational fear. Of course, public concern over the effects or impact of new technologies is cyclical. Throughout history, new ways of doing things—particularly new ways of communicating—have cut through society like a double-edged sword, providing the means for both liberation and empowerment, but also threatening to sever society's "power elites" from their perceived entitlements. As such, a disquieting tension sometimes exists among a mélange of competing forces racing toward some as-yet indistinct pot of gold. Is the Digital Age good for society? There is no yes or no answer to this question. The answer is always yes AND no.

The Digital Age is not just about the big players in the digital media industry, such as America Online, Time Warner, Amazon.com, and Microsoft. Mainly, it is about ordinary people using digital media for chatting with friends, doing research for school or personal interests, checking out cool Web sites, keeping up to date on the latest news, buying a book or CD, watching and listening to a seminar halfway across the country, and participating in any number of other digital events and activities.

In times of social change and adaptation, it is best to arm oneself with the knowledge for creating options. There have been many communication "revolutions" in the past because of technologies such as the printing press, telegraph, telephone, radio, and television. Each of these technologies has had a dramatic and irreversible impact on media and society. They have invariably created tensions among the competing forces of government, market, and civil society. Never in the history of humankind, however, has there ever been anything like the web of digital media currently wrapping itself around the globe. Never before have so many people had the capacity not only to *connect* to such an extensive and powerful network of resources (people, machines, and content) but also to engage as interactively with those resources as a *communicator* and *producer* of information. It is an opportunity like none other in history.

This is an excellent time for people, young or old, to explore the world around them in ways never before possible. This book helps lay the foundation for understanding the constantly emerging and changing digital media environment. It is hoped that astute students of digital media will be able to make more informed decisions about their future academic, personal, and professional lives armed with this basic knowledge. They should make wise choices in their consumer options, political representation, and civic engagements. Intelligent public deliberation that can influence policy making is most effective with well-informed constituencies and electorates.

The pitfalls of digital media should not be ignored. Along with the capacity to transmit news and information in greater volumes and higher speeds than ever before comes more opportunities for computer-mediated antisocial activities such as rumor mongering, exchanging and perpetuating false information, and a kind of social and civic malaise due to so-called information overload. Some critics argue that in the digital media environment, the important stuff gets lost in an overflowing sea of petty irrelevances.

The misuse of digital media can have serious consequences for innocent people's reputations, bank accounts, and livelihood. For example, on August 25, 2000, a bogus press release with false information about the fiber-optic company Emulex led to a rash of incorrect news stories electronically circulated throughout the world. The stock value plunged more than 60 percent in as little as 15 minutes because of these false reports. The company recovered when the hoax was revealed, but by that time many investors had sold their shares at steep losses as they tried to bail out of what they thought was an unstable company. Illegal and unethical activities on the Internet have been increasing and can have serious consequences. It is not always clear how best to deal with these activities. More laws? Technological solutions? Hands-off approach?

Students who specialize in digital media in college have the potential to work in many different kinds of jobs. For example, with a degree in traditional journalism and knowledge in digital media, one could work in the new media divisions of large or small media organizations. Almost all forms of professional communications (journalism, public relations, advertising, broadcasting) offer services in the digital media these days. But even outside of professional communications jobs, there is a need for people who are skilled and knowledgeable in digital media—for instance, retail outlets (both small and large businesses), nonprofit organizations, academic institutions, religious groups, government offices (an enormous amount of government information and services is available online), and so forth. Many people have started their own businesses online, their own informal news and information sites, and their own hobby sites. By the end of this book, you will have acquired a proficiency in discussing a wide range of issues and specific details relating to digital media. Having acquired this basic "digital literacy," you will be better prepared to continue the learning process at more advanced levels.

Chapter Summaries

Media and Society in the Digital Age is organized into 11 chapters, the sum of which provide a comprehensive, descriptive, and analytical view of the world in this period of profound social and technological change. Chapter 1 provides an overview of media and society in the Digital Age, focusing in large part on rapid growth of digital media, such as the Internet, World Wide Web, and developments in satellite broadcasting. This chapter also examines the history of mass media, their role in society, and the cyclical public concerns that have emerged about their impact on groups and individuals.

Chapter 2 examines the deployment of an information infrastructure and the development of digital networks in the United States and beyond. The growth of computer networks, satellite technology, and broadband channels of communication has laid the groundwork for twenty-first–century media.The ease with which news, information, entertainment, and financial data can travel, or be stored, retrieved, and manipulated in various ways, is unprecedented. This chapter will examine the decades-old deployment of national and international networking, beginning with some of the earliest examples of nonelectronic networks—such as roads and maritime trade routes, human couriers, and semaphores—to developments in the early nineteenth century after experiments with electrical signal transmission (first through wires, and then through the air) gave rise to technologies such as the telegraph, telephone, and wireless telegraphy (or early radio). A discussion of the early web of communications prior to World War I will provide the foundation for succeeding layers of networking to be described—namely, broadcast radio, television, satellites, and the Internet.

Chapter 3 will look more specifically at changes affecting traditional mass media, such as books, newspapers, magazines, radio, and television. Whereas the previous chapter discusses mainly the conduits or channels of communication, both old and new, this chapter looks at content—primarily the news, information, and entertainment that travel over the increasingly complex national, international, and global communication and information networks. As books, magazines, and pamphlets provided the means for ideas to disseminate to ever larger numbers of readers, and as literacy increased among the burgeoning middle class in a more "enlightened" eighteenth-century Europe and America, the media's role as a political and social "change agent" should be examined. Radio and television in the twentieth century further affected the cohesion of an increasingly dispersed (and diverse) population in the United States. Are digital media precursors to a more cohesive or a more fragmented society? There is a complex answer to this deceptively simple question. Digital media will likely impact society in ways similar to how print and broadcast technologies affected society, allowing for the spread of news and information over greater distances in larger volume with accelerated speed from one communication "era" to the next. By the same token, these same technologies can also be used to circulate hate, disinformation (propaganda), and criminal communications farther, faster, and in larger doses. There are advantages and disadvantages to technological progress.

Chapter 4 is perhaps the most technical of the chapters and deals with the "nuts and bolts" of digital media. It is a primer about digitization, convergence, and multimedia. This chapter describes computer language—bits and bytes—and how text, graphics, photos, sound, and video can be converted into computer-readable code that is easily stored, shared, transmitted, and manipulated.

Chapter 5 explores the penetration of digital media into our everyday lives. The *digital revolution* is really a misnomer—it suggests that changes in the communications environment have occurred practically overnight, suddenly and furiously. In fact, a series of quiet digital revolutions have occurred over past decades, but especially since the early 1980s. In one form or another since then, computers have gradually slipped into our everyday lives—at the bank, at the grocery store, at the library, at the medical center, at government offices, even at the used car lot. Many common household appliances have computer chips in them. This chapter shows how pervasive digital media have already become.

Chapter 6 takes a penetrating look at an increasingly important topic in contemporary global society: media economics, e-commerce, and the digital economy. The private sector is driving the development of digital media, and big bucks are being made and changing hands. Transnational multimedia conglomerates have taken risks and invested heavily in digital media, but many "start-ups" have also profited handsomely (if sometimes only in stock value) from a global high-tech feeding frenzy. In recent years, electronic commerce has gained widespread attention in business circles. The ability to make financial transactions on the Web and engage in an exchange of goods and services previously confined to the "real world" has opened a whole new playing field for those interested in capitalizing on this as-of-yet "undertapped" resource. No one knows for sure whether or what fortunes wait behind digital media ventures, but many have already taken the plunge.

Chapter 7 discusses government policy, planning, and regulation. As with every new technology throughout history, the digital media are coming under the scrutiny of government and special-interest groups that believe it is important to regulate some aspects of communication and information flow. Sometimes government regulation of information industries may be contentious and even coercive; other times regulation is simply motivated by the need for coordination, cooperation, and fairly mundane administrative tasks. The formation of the International Telegraph Union (ITU) in the 1860s, for example, harmonized tariffs among participating nation-states and led to the creation of technical standards. This chapter describes the development and impact of government planning and policy making in respect to emerging technologies throughout history.

Chapter 8 discusses the interrelationship among government, market, and civil society, and how these three social forces compete with and complement each other in a digital media environment. This chapter examines the role of each of these sectors and how problems can occur when their respective interests fall out of balance. In a democracy, however, these forces should provide a check on each other. They can even work toward common development goals under certain ideal conditions. Examples will be given of how a holistic integration of government, market, and civil society interests can facilitate a productive, functional, and equitable digital media environment.

Chapter 9 turns the reader's attention to the process of inquiry and discovery. Communication research has always been an important part of the social sciences, even before communications was a formal field of study in U.S. universities. Scholars such as Robert Park and John Dewey, for example, were asking questions about the media's (in those days, mainly newspapers) role in creating and maintaining a sense of community among diverse individuals living within common borders. Another line of inquiry had to do with the impact of respective media on audience perceptions and behaviors—especially (but not exclusively) relating to children. How can traditional media research be applied to digital media? What do we need to learn and understand? Are there concepts and theories that can help us better explain and understand the phenomena of digital media?

Chapter 10 exposes the reader to a critical view of digital media. There have been many critics of the "old media"—Ben Bagdikian, Vincent Mosco, Todd Gitlin, Neil Postman, and others—and there is an emerging brood of digital media critics as well, including Clifford Stoll *(Silicon Snake Oil)*, David Shenk *(Data Smog)*, and Michael Noll *(Highway of Dreams)*. What criticisms do they have of digital media? This chapter takes a serious look at these criticisms and invites the reader to decide whether they need to be heeded.

Chapter 11 turns an eye to the future. No one knows for sure what the digital media environment will look like 20, 10, or even 5 years down the road.What we as a society do today—the decisions we make and the directions we map out—will pave the way for continued evolution in the digital media environment. What might that future look like?Where are we heading?What cultural, social, technological, economic and political factors may impact the communications landscape of tomorrow, and vice-versa?This chapter provides a hopeful but well-reasoned "guided tour" through some potential scenarios along the digital media landscape.

Acknowledgments

To personally thank all the people who have contributed to the development of this book would require a chapter unto itself. Many people in my life—teachers, colleagues, friends, family members, students, librarians, and others—have provided considerable knowledge, inspiration, and moral support. To issue a blanket "thank you" seems so insufficient, but I hope they all know how genuinely appreciative I am for their generosity and kindness.

A number of my colleagues at The Freedom Forum Media Studies Center in New York City, at the University of Washington in Seattle, at the University of Hawaii, and at the East-West Center in Honolulu were particularly instrumental in helping to shape my intellectual and professional interests. I deeply value their friendship and guidance. Their contributions have found their way into this book in different ways.

The professional team at Addison Wesley Longman and Allyn and Bacon have made the experience of writing a textbook gratifying. Molly Taylor saw this book to its completion, and Michael Greer worked with me on it during its early development. Others who have been helpful at different stages along the way include, in alphabetical order, Jennifer Becker, Erika Berg, Karon Bowers, Michael Kish, Priscilla Mcgeehon, and Rich Wohl. Lynda Griffiths of TKM Productions provided detailed and thoughtful editing that considerably improved this book's final copy. I also thank the following reviewers for their suggestions and comments: Katrina Bell-Jordan, Northeastern Illinois University; Michael Brown, University of Wyoming; Richard Craig, San Jose State University; Eric Reed, Owens Community College; Darrell L. Roe, Marist College; and Shelly Wright, State University of New York at New Paltz.

Finally, my parents, Ralph and Barbara, and all of my family deserve special mention for their unwavering support over the years.

Introduction

Change and Uncertainty

I learned of the terrorist attacks on the World Trade Center from the Internet. One of the first things I do before beginning my day, even before reading a newspaper or watching television, is to scan a variety of online news sites to help me keep up with current events. On the morning of September 11, 2001, the first news site I visited was CNN Interactive. The headline on this breaking news story said that a plane had crashed into one of the towers of the World Trade Center in New York City.

As a former resident of Manhattan, my heart jumped. I ran downstairs to the television and turned on the morning news. All of the anchors were talking about the situation, not so much as a terrorist attack at first, but as a tragic event that their news crews were still investigating. Soon, video footage of the crash was being aired as I—and tens of millions of other people—watched in horror and disbelief the scenes that have since been repeated over and over again.

After a while, I ran back upstairs to the computer and e-mailed a friend who worked in downtown Manhattan. "I can't believe what I'm seeing on television," I wrote. "Are you okay?"

She wrote back that she was okay, but she couldn't get in touch by phone with her sister and brother-in-law, both of whom also worked in Manhattan. The phones were not working. She asked me to call them from Seattle, where I was living, to see if I could get in touch with them, and to e-mail her back if I was successful in reaching them. (I tried but could not reach her sister or brother-in-law. To everyone's relief, they were later found to be safe.)

In a short period of time on that fateful morning, a number of communication technologies, old and new, mass and personal, wired and wireless, were put into action all over the nation to (1) collect and disseminate information about what was happening in our environment, (2) help reduce fear and uncertainty and (3) communicate with people on a personal level as well as on a one-to-many level. Electronic networks of all kinds were being used for national and global communications.

The Internet was not an exclusive source of communication and information but a *complementary* source, along with traditional mass media and personal communication devices such as the "old-fashioned" telephone. In the days that followed the attack, the public learned that cell phones played an important role in personal communications as people called their loved ones from the World Trade Center (before it collapsed), from the hijacked airplanes, and from other places where conventional telephone connectivity was impossible.

The nation's and the world's communication networks were doing what such networks have done even in their most primitive manifestations: They shunted news and information over long distances so that people could think, feel, and respond in ways they believed were most appropriate. But they also facilitated a conceptual space for deliberation, mourning, anger, understanding, and confusion. People used digital communication tools such as electronic discussion groups or Web sites to post their thoughts on what happened, argue or empathize with other people, and just vent their frustrations. Parts of the Internet were like virtual communication watering holes for people who just had to "talk" to other people.

The mass media and personal media devices allow people to learn and to communicate beyond face-to-face, real-time interactions. Insofar as they lead to a more informed and empowered society, they have socially constructive functions and can enhance community building and democracy. Not all uses of communication technologies are benevolent, however. Sometimes the uses can be hateful and destructive, such as in racist and homophobic Web sites. After the September 11 attacks, some security and terrorist experts have claimed that secret, hidden messages can be embedded in the most innocuous-looking content on the Web. Called *steganography*, the process would invoke a centuries-old practice of hiding messages that only the intended receiver or receivers know how to decode. These messages could provide instructions and details for future terrorist attacks. It is not clear whether steganography has actually been used for such purposes with regard to the Internet, but researchers have been studying the Internet for hidden codes, and the National Security Council has warned media organizations about its possible use by terrorists.[1]

The concerns about criminal and terroristic use of the Internet and other communication networks in the United States are understandable, but some people are also concerned that law enforcement officials, with the public's fearful consent, may overstep their authority and threaten civil liberties as they investigate potential security breaches. The use of wiretapping and other surveillance technologies designed to monitor electronic communications tends to increase during times of war. Modern-day surveillance technologies with names like Echelon and Carnivore allow law enforcement agencies to intercept and examine messages sent over communication networks. Their use has not been without controversy.

Clearly, digital media in the twenty-first century entails a complex web of technologies, practices, policies, controversies, relationships, and economic considerations. Writing about this emerging media environment is like walking a tightrope in anticipation of an earthquake: You have to be confident that you are up to the task, but you also have to be prepared for any number of unexpected circumstances that could throw you off course. That is essentially what happened between the time I started writing this book and when I completed it. As the 1990s drew to a close and the year 2000 sprung forth with much fanfare and anticipation, several significant events provided some unmistakable clues that the world of digital media was finally coming of age. Hundreds of Internet-based businesses—"dot-coms"—were being launched with the help of high-tech venture capitalists. Savvy investors eagerly awaited the announcement of initial public offerings (IPOs) for new companies whose paper value was based more on hype than a realistic business model.

Early in 2001, however, reality set in. The stock market experienced a series of euphemistic "corrections," and the stock value of many technology start-ups, as well as more established companies such as Microsoft, Intel, and Cisco, took a rapid nosedive. According to an article in the *New York Times*, more than 65,000 employees were laid off from Internet-related companies between December 1999 and March 2001.[2] The downturn was expected to continue as the dot-com industry struggled to find viability. Not surprisingly, the stock market suffered serious losses after the terrorist attacks of September 11, and the U.S. economy struggled to get back on its feet.

Despite the uncertainty surrounding dot-coms, few people were predicting the end of Internet-related businesses. It had become axiomatic that we were living in the Digital Age. The Internet and digital media were here to stay, and there was no turning back: Change was at hand, for better or worse.

The mass media—both news and entertainment—can serve as harbingers of change when they focus on issues, people, or events that signal a shift in traditional thinking. Throughout the 1990s, the news media, seemingly *en masse*, had been heralding the promises and perils of the impending "digital revolution." This term was plastered across magazine covers and was the subject of countless news stories in other media. The suggestion of cataclysmic change in the communications environment tied in nicely to the public's anticipation of the new millennium. Indeed, in the transition between 1999 and 2000, *Time* magazine took the bold step of choosing a rather unlikely individual to grace the cover of its annual Person of Year issue. The selection would prove to be both prescient and ironic.

Time*'s Person of the Year*

Every year since 1927, *Time* magazine has announced its Person of the Year in the first issue of January. Rarely has the magazine's announcement been a total surprise. The usual suspects tend to take home the prize: U.S. Presidents, peacemakers, leaders of revolutions of one kind or another around the world.

At the end of 1999, *Time* put a relatively obscure but visionary entrepreneur named Jeff Bezos on the cover of its Person of the Year issue. "*Who?*" many pundits joked at the time. Others who heard the news asked the same question but were serious. People may not have known who the then 35-year-old Bezos (one of the youngest ever to have won the award) was, but almost everyone has heard of the company he founded: Amazon.com.

The year 1999 was one of those rare occasions when *Time* decided to take a chance and go out on a limb with its Person of the Year selection. It had many candidates from which to choose: Alan Greenspan was doing a great job chairing the Federal Reserve; China's premier Jiang Zemin brought his country closer to global integration than any of his Communist forbears; Steve Jobs resuscitated a moribund Apple Computer company; and the country's first female Secretary of State, Madeleine Albright, received high marks for her adept foreign diplomacy and would have helped even out the gender gap to boot. Joe DiMaggio and John F. Kennedy, Jr., were logical posthumous candidates. But none of these name-recognition all-stars ultimately made the final cut. The comparatively nondescript Bezos took home the honors.

Not everyone agrees that the magazine made the right decision. In fact, an online (read: unscientific, but still fun to quote) poll conducted on the magazine's Web site soon after the Person of the Year was announced showed a solid majority of respondents disagreeing with the decision. But in many respects, Bezos was an ideal choice, not necessarily because of who he was at the time, but because of what he represented. Jeffrey Preston Bezos was emblematic of a sea-change in human consciousness, a phenomenon that will probably permeate all of global society in one form or another. On a fateful day in 1994, when he found a Web site claiming that the Internet was growing at a rate of 2,300 percent a year, he began to hone in on an idea that he believed would revolutionize book selling specifically and the world of retail more generally. He saw both opportunity and a new way of doing things, even as many around him—like investors—shook their heads. In selecting Bezos as Person of the Year, the magazine recognized that through the ideas and initiative of people *like* him, the engines of social, economic, technological, political, and cultural change would find the energy and momentum to transform the world as we currently know it.

The magazine's choice was surprising, but if the past is any indication, *Time*'s least expected and perhaps most perplexing choices for Person of the Year have also been its most refreshingly prescient. The 1982 Person of the Year, for example, was astonishing for what it was *not*—a person. The magazine declared the computer as its Machine of the Year. In explaining its decision, it cited a poll (which the magazine commissioned and which used scientific methods for public opinion polling) that showed "nearly 80% of Americans expect that in the fairly near future, home computers will be as commonplace as television sets or dishwashers."[3] There was also concern: Although respondents perceived dangers of unemployment and dehumanization as a result of more computers, many of them felt "the computer revolution will ultimately raise production and therefore living standards (67%), and that it will improve the quality of their children's education (68%)."[4]

Clearly, that year the magazine had its finger on the pulse of impending change in American society.

What it may not have foreseen either in 1982 or in 1999 was how close to home that change would strike. In the early weeks of the year 2000, it was announced that the magazine's parent company, the media conglomerate Time Warner, was being bought out by America Online (AOL), the largest and most well known online company but still remarkably young and inexperienced to be buying a global media and entertainment behemoth.

An Event and a Nonevent

The AOL/Time Warner merger was the first big digital media news of the year 2000. Well, actually it was the second. The first was the reporting of a nonevent: "Armageddon" (technologically based and otherwise) had not occurred during the transition from 1999 to 2000. After all the hype about potential Y2K disasters—airplanes flying without control tower guidance, consumer utilities malfunctioning, financial transactions going haywire, and people holed up in their make-shift bomb shelters with enough food and water to last them as long their ammunition held out—the quick journalistic wrap-up in the new year was that everything was fine and life could now go on as usual. The term *Y2K* became more synonymous with the actual year that it stands for rather than with the disasters everyone was pretending they never believed would occur anyway.

But the second big story was the Time Warner buyout. The announcement was headline news, not only because it would create the largest multimedia conglomerate in the world by far but also because it represented a marriage—or a convergence—between a relative newcomer in the communications industry (AOL) and a formidable media powerhouse (Time Warner) whose roots were well established in traditional forms of media such as magazines, books, films, and recorded music. And it was the newcomer buying out the old-timer. Many wondered how a company founded in 1985 by a young Hawaii-born entrepreneur named Steve Case could buy the largest media conglomerate in the world for $165 billion. For those last holdouts who predicted "new media" companies were merely a passing fad, perhaps it was time to start taking the Internet-based communications industry seriously. Slowly but surely, digital media were being integrated into traditional communication infrastructures, and the walls separating previously discrete forms of media (e.g., books, newspapers, magazines, radio, and television) were blurring. Much enthusiasm and optimism surrounded the emerging digital media environment, largely for its impact on the economy, but also for the new model it provided as a distributor of news and information, and for its potential to develop meaningful civic networks.

The Digital Age Is Here

For years, evidence of an impending digital millennium was growing. URLs, the Web "addresses" used to find a particular Web site on the World Wide Web (WWW), began popping up all over the place—in newspapers and magazines, on television, on billboards, on the sides of buses, on storefronts, even on paper napkins and on popsicle sticks by the mid-1990s. Traditional news organizations were offering online services. A small number of television stations across the country began broadcasting in digital signals in anticipation of the day when digital television (DTV) would be the norm. Colleges and universities were wiring their classrooms and dormitories for their students to get direct access to the Internet. Financial institutions were encouraging people to "bank online."

Of course, there were those who dismissed all the brouhaha as a passing fad. At the end of 1995, a Montreal, Canada, newspaper writer ruminated on the future of the Internet as a new year approached. "I'm here to tell you," wrote Doug Camilli in a New Year's prediction column, "that 1995 will go down in history, quite far down, as the year before the Internet fad collapsed." He guessed that a history book in the future would say that by mid-1996, for a number of reasons, the Internet would begin to collapse after years of exponential growth like other short-lived attractions in history.[5]

That same year, the (London) *Guardian* cited a study saying that the Internet and home computers were a "fashionable fad" and that the PC boom is "thus illusory."[6] Other journalists used the odd logic that since some information on the Internet was unreliable (e.g., the exchange of conspiracy theories, gossip-mongering, half-truths), the *whole* medium should be distrusted, even abandoned. Interestingly, they rarely applied the same standard to contemporary traditional news media, which could hardly be called uniformly responsible and trustworthy.

It is not surprising that some of the most vitriolic comments about the Internet have come from those in the traditional news media and other long-established social institutions. The proposition that the public would soon be able to bypass the journalist and go directly to the source threatened the livelihood not only of individual journalists but of the profession as a whole. It was the fifteenth-century printing press versus the Catholic Church being played out all over again in the 1990s. The root of the fear, however, burrowed much deeper than the superficial issue of job security. It was and continues to be grounded in the prospect of a monumental shift in power relations. The traditional institutions of knowledge production, political authority, and legitimacy, with walls of professionalization and narrow access surrounding their historical domains, were being infiltrated by a bunch of young upstarts with computers.

Needless to say, by the year 2000, there was little talk of "fads" and "illusions." The Digital Age is upon us, and society is feeling its impact. The used car salesman, the retail merchant, the travel agent, the college professor, the librarian, the stock broker, the insurance provider, the grocer, the pharmacist, the bookseller, the political activist, the hobbyist, and, yes, the journalist and other media professionals, all face threats and opportunities in the Digital Age. That persistent adage—"Adapt or die"—has never rung more true than in today's digital media environment. Sometimes change has come without much fanfare or resistance. Most libraries today, for example, reflect a completely overhauled cataloging and circulation process from two decades ago—the result of a gradual but comprehensive conversion to automated systems. People learned to move from card catalogs to online catalogs, and to access an increasing number of documents online in full text. Other changes have taken more time. There are still doubts about the security of financial transactions online, about the ability to ensure privacy of personal information, about the efficacy of online advertising, about the credibility of computer-mediated distance learning programs, and so forth.

But changes are clearly occurring. Over the past few years, record numbers of Internet-based companies have come and gone. Most traditional news media organizations also have online divisions of varying sizes and abilities. Yet, at the same time, with very few exceptions, profits have been evasive and the public's appetite for digital products and services has been difficult to forecast. The sharp rise and fall of the dot-com economy provided a clear signal that many people and companies are still struggling to build a viable, sustainable and socially integrated digital infrastructure.

The Real Digital Age: Dot-Com Maturity

For a period of time around the turn of the millennium, it seemed that digital media companies and all their dot-com cousins could do no wrong. They were waltzing across that tightrope in confident droves, often transitioning from obscurity to rapid financial success stories seemingly overnight. Dot-com stocks went through the roof, and only a relatively few sobering voices could be heard cautioning the public about all the media hype and overvaluation of these emerging Internet-oriented businesses.

A large number of companies—including traditional media organizations—joined in the Internet frenzy, literally banking that their new Internet divisions were going to bring newfound wealth and glory to their existing operations.

"This seemed like a total no-brainer back in 1999, those dear departed days when Internet stocks were riding high, when Priceline.com was valued at more than the entire airline industry and Yahoo was worth more than Disney," wrote Allan Sloan in the *Washington Post.* "It seemed that anyone who could spell 'Internet' would become an overnight billionaire, while mainline corporate America still had to produce profits and cash flow to get stock prices up."[7]

Then came the earthquake. Even before 2000 came to an end, dot-coms began falling like dominoes. The transition from 2000 to 2001 was a rocky one. Rather than overnight success stories, the news media began reporting about plummeting stock values, massive corporate layoffs, company closings, projected quarterly earnings that were going unmet, leveragings, repositionings, bankruptcies, and the ubiquitous practice of "downsizing." In the first quarter of 2001, both America Online and Amazon.com had announced that they would proceed with layoffs that, combined, would affect thousands of their employees. Many smaller companies disappeared just as quickly as they sprouted out of nowhere. Within the span of about one year, *Time* magazine's Person of the Year Jeff Bezos went from making front-page news for his pioneering success as an online bookseller to explaining why it was in his company's best interest to layoff 1,300 employees in one fell swoop.

Nationally, Wall Street's anxiety over Internet-based businesses was reflected in the plunging numbers on Nasdaq, a composite index composed primarily of technology stocks. For example, petstore.com, before it went out of business entirely, saw its (at one time promising) stock lose 98 percent of its value. In July 2001, Webvan, the online grocer whose stock at one time sold for more than $30 per share, turned off its engines and filed for bankruptcy about a year after taking over its competitor, NetGrocer. In its final weeks, Webvan's stock was worth pennies per share. Less extreme cases saw companies surviving but with stock values that were half or less of their all-time high. Even large companies such as Microsoft were not spared. Founder Bill Gates is said to have lost billions of dollars in a matter of days (although at last writing he and his family were never in danger of being thrown into the streets). Others did not fare as well. Tens of thousands lost jobs and, hence, their livelihoods. Others borrowed large sums of money against their stock options and, when the stock market plunged, they were literally indebted to their banks.

One company that has come to symbolize the rapid rise and fall of digital media experiments in the 1990s is Yahoo, which describes itself as a "global Internet communications, commerce and media company" and claims to provide services to nearly 200 million people. At one time valued at $70–100 billion, the company has seen a precipitous fall in its stock price from a high of $250 to about $16 a share in April 2001. Although the company is still afloat and struggling, like so many other Internet companies, to regain some of its lost glory, it is often pointed to as an example of how volatile and diffident the Internet business really is. Within the span of a year or so, the perceptions of many investors about the market viability of Internet businesses have changed markedly.

The Digital Age turned out to be not as predictable as some might have thought. Even though the number of people who had access to the World Wide Web from work and home grew dramatically in the 1990s (from 22 million users in North America in 1995 to more than 90 million in 1999),[8] research on behavior patterns and effects of users were sketchy. No one really knew what the potential of the Internet was, although many banked on its success as a vehicle for e-commerce, online advertising, paid subscription services and other commercial uses. Many of these uses have yet to prove viable. More certain is the dangerous unpredictability of the Internet. Fortunes can literally be made and lost in a matter of hours thanks to online technology (e.g., through day trading); great ideas can soar and then disintegrate; famous people can be at the top of the world one year and then fighting for their companies' survival the next. The speed at which communication and commerce occur locally, nationally, and globally has accelerated in this increasingly interdependent, interconnected world in both positive and negative ways. For example, news and information travel faster, but then so do gossip and rumor. The next chapter provides a broad overview of factors that have contributed to the development of a digital media environment.

Endnotes

1. Scott Shane, "Code Experts Say bin Laden Could Have Hidden Message," *Baltimore Sun*, October 12, 2001, p. A-1.
2. Sam Lubell, "No Pink Slip. You're Just Dot-Gone," *New York Times*, March 18, 2001, p. 2. The article quoted data from Chicago placement firm, Challenger, Gray and Christmas.
3. Otto Friedrich, "The Computer Moves In," *Time*. January 3, 1983, p. 14.
4. Ibid.
5. Doug Camilli, "The Decline of the Internet... and Other 1995 Phenomena," *(Montreal) Gazette*, December 30, 1995, p. D-1.
6. Nicholas Bannister, "Novelty of the Net Begins to Pall as Public Prefers Playing Games," *(London) Guardian*, November 1, 1995, p. 21.
7. Allan Sloan, "Demise of NBCi Shows the Dangers of Trying to Create Internet Wampum," *Washington Post*, April 17, 2001, p. E-03.
8. http://www.commerce.net/research/stats/wwwpop.html. Statistics on Internet use are inexact and debatable. They are most valuable as estimations and for visualizing trends. As of June 2001, Nielsen//NetRatings.com estimated that 167.1 million people had home Internet access. See: http://www.nielsen-netratings.com/.

1

An Overview of Digital Media

Factors Leading to a Digital Media Environment

The emergent digital media environment affects all aspects of society, from news and information to commerce, education, entertainment, health care, government, politics, love, sex, death, crime, deviance, religion, international relations, and civil society. Before getting into specifics, however, this chapter will provide you with a "user-friendly" overview about what digital media are, how the digital media environment came to be, and why this topic is important to your life.

There was a time when news traveled only as fast as human feet could carry it. Human couriers were the medium of transmission, and one's ability to run long distances could be both a career advantage and an occupational hazard. Legend has it that in 500 B.C., a soldier ran more than 22 miles to Athens to announce Greek victory at the Battle of Marathon, and then promptly dropped dead from exhaustion. The combination of human ingenuity and necessity gave rise to inventions—at first crude and rudimentary—that allowed information to be conveyed over long distances without direct human agency. Smoke signals, carrier pigeons, firegrams, a series ofvisual telegraph towers (also called *semaphores*), the electric telegraph, the telephone, and broadcast technologies were some of the progressively innovative ways that humans discovered how to send messages—sometimes as code—over long distances in a fraction of the time it would take to carry that same message even by the fastest runner. The concept of *telecommunications* (*tele* means "a great distance") was born.

Today, telecommunications implies the use of *electronic* technologies for communication over long distances. Since the early 1800s, when inventors experimented with sending messages in the form of electrical impulses over metal wires, the use of coding and decoding devices to send and receive messages has resulted in ever more sophisticated telecommunication technologies. These technologies were eventually deployed as domestic telecommunication networks, and then evolved into more complex networks across national borders. Today, they comprise a global information infrastructure.

The convergence of computer networks, satellites, telephone connections, and other technologies with traditional mass media, such as newspapers, magazines, radio, and television, has resulted in what some refer to as a *digital revolution*. It is called this because information—whether embodied as text, sound, moving images, photographs, or graphics—is converted to a similar "language" (coded as binary digits, in this case "1"s and "0"s) that can be quickly read and exchanged by computers, and then decoded, again by computers, and presented in a form that people can comprehend. This conversion process is known as *digitization.*

Digitization, however, is just one of a number of important factors that have led to the emergence of "digital media" in North America—that is, media that tend to be digital, computer driven, interactive, and, in many cases, able to more specifically serve the needs of an information seeker than the traditional media. Because *new media* is a time-sensitive designation (at some point, this "new" media will no longer be new), the term *digital media* will be used instead. Digital media also permits vast and rapid dissemination of information. Compared to "old" (or traditional) media, such as books, newspapers, magazines, radio, and television, the digital media are less constrained by time (in the case of broadcast media) and space (in the case of print media). The digital media have a global reach.

Global electronic linking mechanisms, such as the Internet, the World Wide Web, and commercial online services, allow traditional news organizations, large and small businesses, nonprofit organizations, nongovernmental organizations, hobbyists, school children, college professors, communities of all sorts and sizes, and a wide range of other people and organizations a way to provide news and information to a virtually worldwide audience or, conversely, to a relatively small, highly specific, and targeted audience.

What has led to the emergence of this digital media environment? Certainly not any one single thing. Rather, a number of factors, working together, have changed the communication landscape in dramatic and revolutionary ways. The literature about digital media is voluminous. Much has already been written about the digital revolution and emergent communication and information technologies, especially the Internet and World Wide Web, from every possible perspective—economic, political, cultural, technological, even moral.

Some of the more seminal and comprehensive works in this area include books by Benedikt (1991), Compaine (1984, 1988), Fidler (1997), Forester (1987), Gilder (1989, 1992), Malamud (1992), Mirabito (1994), Negroponte (1995), and Pavlik (1996). More wary yet informed perspectives are offered by Mosco (1989), Noll (1997), Stoll (1995), and others. Literally hundreds, if not thousands, of articles supplement this list. Although it is outside the scope of this chapter to provide an exhaustive account of the history of digital media (these can be found elsewhere), it is possible and important to provide a thumbnail sketch of the multitudinous factors that have resulted (or are resulting now) in transforming media and society in the Digital Age.

The following summary is a distillation of numerous sources that have attempted to detail, in one way or another, some aspect of digital media. These factors represent 10 major categories[1] of reasons for the emergence of digital media. They are provided here as background and "scaffolding" for further discussion of digital media and are meant to be, for now, a simple overview. More detailed discussion of many of these factors will occur in subsequent chapters.

In a nutshell, the 10 factors[2] are:

1. Increased digitization of information (text, audio, video, photos, and graphics)
2. Growth and mass penetration of powerful personal computers
3. Development of user-friendly interfaces and miniaturization of computer hardware
4. Development of networking software and hardware
5. Federal government support for the building of an interoperable global information infrastructure and the revision of antiquated telecommunications laws
6. Corporate consolidations in the media and telecommunications industries
7. Technological convergences
8. Broader bandwidth capabilities and compression technologies
9. Diffusion of computer technology in many sectors of daily life, including education (and libraries), business, government, health, and civil communication
10. Demonstrated market demand for news, information, and entertainment

Digitization

When information is digitized—that is, converted into a logical series of 1s and 0s that a computer can read and change into a form that humans can comprehend—it makes no difference whether the information is text, sound, moving pictures, photographs, drawings, and so forth. "It's all digital!" as the saying goes; all of these forms of information are simply specific combinations of 1s and 0s. For example, at a simplistic level, the letter A can be coded as *01000001.* Each number (the 1 or the 0) represents the smallest unit of digital code, or a *bit* (binary digit). Eight bits equal a *byte.* Depending on the complexity of the information (e.g., a simple alphanumeric character versus a complex array of sound and video), the amount of digital code needed to digitize information can vary widely. This digital information can then be transmitted via a number of different methods, through fiber optics, radio waves, telephone wire, coaxial cable, or a combination of some or all of these.

Computer-readable code is the basis of digital communication, everything from e-mail to Web surfing to e-commerce. The digital revolution assumes a society in which considerable amounts of the research, storage, production, presentation, and dissemination of information involve digital technologies and digital code.

Consumer Adoption of Personal Computers

The number of personal computers in U.S. homes pales by comparison to technologies such as the television and radio, but it has been rising rather rapidly since the early 1980s. In 1983, only 7 percent of U.S. homes had a personal computer. By the year 2000, more than half of all U.S. households had a computer. And in July 2001, some reports had that rate of penetration as high as 63 percent.[3] Of course, simply knowing penetration rates does not tell us *how* people are using their computer, but the point is that personal computers, which before the 1980s were virtually nonexistent in the average American home, have made inroads into the homes of ordinary people, particularly those with children.

"Friendlier" Technology

In the old days, computers were large, intimidating, and unfriendly. The early computers developed just before, during, and after World War II were physical monstrosities by today's standards. The first U.S.-built, all-electronic computer, known as ENIAC (for Electronic Numerical Integrator and Calculator), took up 1,500 square feet of floor space and consisted of more than 18,000 vacuum tubes. It was tangled in a mess of cables and switches and weighed 60,000 pounds.

With the invention of the transistor and other innovations, computers did become smaller, but for decades after the end of World War II, they were still relatively large, highly complicated for the nonexpert, and geared for use by government and big business. In the mid-1970s, the first truly "personal" computer appeared in the form of the MITS Altair. Probably unrecognizable as a computer by today's general user, the Altair system could hardly be described as "friendly," although its size was certainly an improvement over many of its predecessors and paved the way for the friendlier computers to follow.

Thanks in large part to the technical expertise and vision of such developers as Steve Jobs and Steve Wozniak, cofounders of Apple Computer (initially a tenuous little start-up company) as well as the more established computer giants such as IBM and Digital, the personal computer industry was destined to boom in the 1980s. Small, easy

Information at Your Fingertips
With a home computer linked to the World Wide Web, people have access to a wide range of information irrespective of geographical location.

to use, and geared for the nonexpert, the personal computer benefited from the development of user-friendly software (i.e., software that helps "hide" or disguise the complexities of computer technology behind an interface that is either somewhat intuitive or fairly easy to learn). The "point-and-click" interface of Apple and then Microsoft and others allowed users to point their cursors at a certain object—a scissors, for example—and then click on that object to initiate a particular command, in this case "to cut." The development of these user-friendly interfaces brought the computer down to the level of the ordinary user who could intuit complex commands rather than rely on knowledge of programming language and logic.

Networking Hardware and Software

When computers were first conceptualized and developed, they were mechanisms for *computing*, not communicating; in fact, prior to World War II, the word *computer* referred not to a machine but to a *person*—one who computes. The earliest mechanical computers were used for adding, subtracting, and other calculations because they were fast and removed human error from the calculation process (although human error can always enter the picture through faulty programming, data entry, and so forth). As computers became more sophisticated and performed a wider range of functions, their potential for communications became apparent. Protocols were developed that allowed computers to "talk" to each other, or to allow people to "talk" to each other *through* computers, and to let people "talk" directly to computers and vice-versa. Perhaps the most famous of these is the Transmission Control Protocol/Internet Protocol (TCP/IP), which allows packets of digital information to traverse multiple networks on their way to their final destination.

Protocols are important in allowing different computer systems to communicate with other (interoperability). When computer systems are interoperable, they can share files, exchange electronic mail, and use software and access data located on a central "server" computer. Local area networks (LANs) connect computers in a physically contained area (an office, for example). Wide area networks (WANs) are like LANs, except, as the name implies, they tend to be much bigger in scope. WANs can connect many different LANs and other related systems together for widespread interconnectivity across offices, campuses, cities, states, and even countries.

The Internet, based on the TCP/IP protocol, facilitates global computer connectivity among similar and dissimilar computer systems. The protocol acts as an effective Esperanto for computers, allowing them to understand and communicate with each other. With an assortment of hardware and software (computer, high-speed modem, phone line or ethernet connection, SLIP [Serial Line Internet Protocol] or PPP [Point-to-Point Protocol] software, etc.), and a growing field of commercial and nonprofit Internet access providers, even the computer neophyte should be able to gain access to the Internet without tremendous frustration or effort. Computer networking made easy has made it easier for computer-mediated communications (CMC) to enter the average household.

Revised Laws and Policies

Early in the first term of the Clinton-Gore administration, both President Clinton and Vice President Gore made clear their support of a National Information Infrastructure (NII), later to be referred to as a Global Information Infrastructure (GII). On September 15, 1993, they released their Agenda for Action, a document calling for the "construction" of a "seamless web of communications networks, computers, databases, and consumer electronics that will put vast amounts of information at users' fingertips." This information infrastructure would be developed by the private sector with appropriate support from the federal government, and interoperability would be a high priority.

The collaboration of government, business, labor, academia, and civil society would be crucial to ensure that this infrastructure would serve many different sectors of U.S. society. The language of a new dawning was formulated as the administration intoned that "America's destiny is linked to our information infrastructure." A few years and many congressional debates later, President Clinton signed the *Federal Telecommunications Act of 1996*, in effect replacing its antiquated predecessor, the Federal Communications Act of 1934. Both the NII and the *Telecommunications Act of 1996* are manifestations of a federal government that is aware of an emerging digital media environment—one that has the potential for putting tremendous amounts of information at users' fingertips and connecting users globally. Both documents also advocate "universal service," however, acknowledging that government oversight of fair access to and distribution of these new technologies remain a distinct probability.

Laws and regulations are constantly being discussed in regard to the Internet. Currently, this is a burgeoning area of public discourse. Politicians and citizens' groups alike are mulling over such controversial issues as online privacy, obscenity, indecency, gambling, crime, commerce, encryption, advertising to children, access to public records, and so forth. The controversial Communications Decency Act (CDA), which was part of the Telecommunications Act of 1996, aimed to regulate what essentially constituted indecent content. This law was struck down as unconstitutional by the U.S. Supreme Court on July 26, 1997. The majority opinion of the Court stated that the law violated the First Amendment's guarantee of freedom of speech. Since then, lawmakers have continued to struggle with the question of how to regulate Internet content and activities. Another law, called the Child Online Protection Act (COPA), similar to the CDA but narrower in scope, was passed by Congress in 1998. In July 2000, a federal court of appeals found the law constitutionally flawed and refused to allow the government to begin enforcing the law.

The attempt to regulate the Internet on many different levels and for a variety of reasons continues. The U.S. Congress and state legislatures and administrative bodies (e.g., public utilities commissions), the Federal Communications Commission, the Federal Trade Commission, the U.S. Department of Commerce, citizens' interest groups, nonprofit organizations, media and telecommunications businesses, and others all contribute to the building of digital media laws and policies.

Conglomeration, Concentration, and Control

The litany of major media and telecommunications mergers just seems to keep getting longer. The Federal Telecommunication Act of 1996, which effectively knocked down a number of regulatory barriers in the media and communications industries and bolstered the hopeful economic potential of digital media services, prompted many large companies to eye each other hungrily. All the big names are involved: Time Warner, America Online, ABC, Disney, Viacom, News Corp., Tele-Communications Inc., MCA, Turner, AT&T, MCI, almost all the Baby Bells, Sony, and Seagrams. Each announced merger seems to top the other. Merger talks between long-distance telephone megalith AT&T and cable giant Telecommunications Inc. (TCI), announced in June 1998, had *Business Week* magazine calling the proposed deal a "turning point in the evolution of communications."[4] The combined Time Warner and America Online conglomerate, finalized in 2001, is today the largest multimedia organization in the world.

The list goes on and will likely continue to grow. Trillions of dollars are transacted each year in mergers and acquisitions, and many of those merged and acquired firms in recent years have included media "properties." Typically, traditional media organizations have attempted to integrate digital media divisions—which might include some aspect of Internet and World Wide Web services, interactive television, video games, and videoconferencing—into their umbrella organization, but as the AOL/Time Warner case illustrates, sometimes the acquisition is the other way around.

There are checks and balances in place to help ensure that a single company does not grow so large as to unfairly eliminate its competition and monopolize the market, although not everyone agrees on the right balance of regulation and deregulation that is necessary to protect the public interest. The media environment is still regulated to some degree by government agencies. For example, some media cross-ownership rules, although substantially relaxed, are still in place. In August 2001, the Federal Communications Commission approved the Tribune Co.'s request to buy television station WTXX WB20-TV in Hartford, Connecticut. But because the Tribune Co. already owned a television station (namely, WTIC Fox61-TV) and a newspaper in the same market, the FCC said that it would have to give up the Fox station within six months. This is in keeping with FCC rules that a media company should not be able to own and operate two television stations in a relatively small media market. (The Tribune Co. owns television and radio stations, newspapers, magazines, and other entities across the nation.) Hence, most media purchases by a conglomerate are still subject to FCC scrutiny and approval.

Convergence and Ubiquity

The buzzword in digital media is *convergence.* So many things seem to be converging in so many ways. Knight-Ridder's (uncompleted) experimental electronic newspaper, the *Tablet,* is a good example of a convergent technology. The format of a newspaper was integrated into the interface of a hand-held computer that made use of hypertext (hierarchically layered) information, audio, moving images that could be played back as many times as the user wished, and online discussion groups. News and information could be customized to the specific interests of the user, unlike traditional print and broadcast media. The project was discontinued by Knight-Ridder, which refocused its research and development on a Web-based delivery model.

A truly ubiquitous communications system is probably still a thing of the distant future, but many developments in digital media suggest that the emergent communications environment is moving in that direction. Eased regulatory barriers are allowing phone companies to offer cable television and cable companies to offer telephone service. A hybrid Web-TV has been on the market but has had unimpressive sales. Advanced television (or digital television, DTV) is expected to be the dominant broadcast medium by 2006. And the Internet—particularly after the introduction of the World Wide Web—is becoming more multimedia as time goes on, functioning as a device for e-mail, Web surfing, word processing, watching movies (using DVD, which is sometimes referred to as digital versatile disk or digital video disc), listening to music CDs, playing interactive games, creating newsletters, and so on.

Some convergences are less apparent than others. Satellite transmission is essential to the operations of the Cable News Network (CNN), *USA Today*, network news, as well as scores of news and information services, but the role that it plays is often not visible or obvious. Convergent technologies will probably become increasingly taken for granted, as interoperable hardware and software permit the seamless transmission of news and information across divergent systems. But like so much else in digital media, the complexity of what is occurring in the background will be hidden by a user-friendly interface. Convergence and ubiquity may well occur unnoticed by those not directly responsible for its realization.

Greater Bandwidth, More Channels

A major obstacle to providing a truly vibrant, ubiquitous multimedia system to a broad sector of the public is lack of bandwidth. In the past, trying to show video through traditional Internet conduits such as telephone lines was like trying to force a tennis ball through a straw: There just wasn't enough room to do it well. This is changing. The deployment of fiber optics, which transfers information in the form of light along thin threads of glass (fibers), can carry the equivalent of 100,000 telephone conversations over a single fiber, obviously a far superior technology to traditional copper wire. But traditional copper wire should not be written off as obsolete. Compression technologies, such as Asymmetrical Digital Subscriber Line (ADSL), compress signals so that more information can pass over traditional telephone lines, making video-on-demand a possible service offered by telephone companies. Another way telephone lines can carry more information is through Integrated Digital Services Network (ISDN). With this service, phone lines carry digital rather than analog signals, permitting much needed high-speed home access to the Internet for those who can afford the service and assuming the service works the way it should.

Other conduits include coaxial cable, which can carry 900 times the information carried on traditional telephone lines. Asynchronous transfer mode (ATM) allows high-speed networking for broadband communications. And digital transmission via satellite and radio waves is certainly part of the digital media future. Teledesic, a venture by Microsoft Corporation's Bill Gates and Craig McCaw (former owner of McCaw Communications, a major provider of cellular telephone service), promises to enmesh the earth's lower orbit with hundreds of refrigerator-sized satellites for global communications networking.

Broader bandwidth capabilities and compression technologies are paving the way for more dynamic content to reach potential consumers of digital media. This increase of bandwidth is essential to the success of digital communication, especially as more audio and video content (so-called bandwidth hogs) take up space on electronic networks. The unacceptable alternative is a sluggish, inconsistent movement of data. Many companies are working to make sure that does not happen. Because of the real possibility of the Internet as we know it getting too congested in the future, a consortium of 180 universities working with government and industry are building the Internet 2, a "leading edge network capability for the national research community."[5] Considerable traffic currently on the Internet would then be siphoned off to Internet 2, further helping relieve bandwidth problems.

An "Information Society"

Computers seem to be everywhere these days—in schools and libraries, stores, banks, hospitals and doctors' clinics, businesses, government offices, video arcades, and increasingly right at home. Sometimes they look like computers, sometimes not, but they're there. Moreover, many people are using computers not only to gain access to news and information but also to "talk" to each other through electronic mail or online discussion groups. Computers and computer networks can link geographically, politically, ethnically, and otherwise diverse people together into "virtual communities" (see Howard Rheingold's [1993] book by the same name) based not on the traditional parameters of physicality and adjacency but on interest and curiosity. Also known as *cyber communities*, these groups often serve as public spheres where uninhibited discussion, debate, and critical inquiry occur. They can also be awash with hate speech, trivialities, and misrepresentations. The onus is on the user to discover a "place" in digital media that suits him or her.

The term *Information Age* preceded *Digital Age* and referred to a society in which information was the basis of its economy, unlike previous eras that revolved more around the agricultural, industrial, or service sectors. In 1977, Marc Porat and Michael Rubin published a series of reports for the U.S. government that talked about the rise of an information economy, in which a predominant part of the U.S. gross national product (GNP) originates with the production, processing, and distribution of information goods and services. Of course since that time, the segment of the workforce dedicated to information production, processing, and distribution has only grown. Whether called the "Post-Industrialist Society," "The Third Wave," or something else, seminal writers such as Daniel Bell, John Naisbitt, and Alvin Toffler perceived decades ago that knowledge and information would transform society in the Information Age in the same dramatic ways that capital and labor transformed societies during the Industrial Revolution. The Digital Age is more a creature of the very last years of the twentieth century, as more and more knowledge and information became linked to digital code and technologies.

Today we live in an Information Society highly dependent on information technologies and electronic networks. Computers have permeated even non–high-tech businesses such as fast-food chains and retail outlets. Cash registers at Starbucks and McDonalds and other stores and restaurants worldwide are electronically networked so that headquarters can keep track of sales and inventory. Record keeping and financial transactions have been automated. News and information race across continents from one computer to another via the Internet or e-mail. Because the conduits through which digital information travels are not always apparent and because digital information (in binary form) is essentially invisible, we may not fully appreciate the amount of information that is "traveling" around us through telephone wires, cables, optical fiber, and radio waves. But think about all the bits and bytes flowing through the world right at this moment in so many ways and for so many reasons. This is, without question, a Digital Age.

Marketability

There are many different opinions about whether digital content (e.g., news, information, entertainment, etc.) is marketable. New ventures on the Internet do not show immediate profits; in fact, only a few kinds of digital media ventures show any direct profits at all. In the future, however, the Internet may be only one manifestation of many digital communication technologies. Some states are talking about constructing information kiosks in public spaces for citizens to access a wide variety of news and information at shopping malls, airports, parks, and so forth. Most libraries and schools already offer free Internet and World Wide Web access. Some communities have set up their own Web sites—the wired neighborhood model, so to speak—for both their own community members and outsiders to access. There is strong consumer demand in North American society for news, information, and entertainment. The question is: Can digital media technologies and services provide what the public wants, the way that it wants it, and generate a reasonable profit or incentive for digital media providers to continue offering and innovating the service well into the future?

Nobody knows for sure, but this question will be asked over and over again, from the corporate boardrooms of AOL/Time Warner and News Corp. to the crowded bedrooms of computer hackers and online hobbyists. What's important is that many people, organizations, and institutions do seem to be investing (sometimes heavily) resources into developing the global information infrastructure in the hope that someday it will prove to be a commercial, educational, or some other kind of success story.

The factors that have been discussed here show how a digital media environment has been evolving in recent years. Digital media have made it possible for many people all over the world to gain access in one form or another (sometimes merely as a service recipient, other times as an actual electronic publisher) to a global telecommunications network and to communicate with others who also have access to the network. The next chapter examines the deployment of an information infrastructure and development of digital networks in the United States and beyond.

Questions for Discussion and Comprehension

1. If you were at a party and someone asked you, "So, you're taking a course about digital media? What is that? What factors are leading to a digital media environment in our society today?" how would you respond? (*Hint:* You should be able to mention at least 10 factors that have led to the current digital media environment.) Also, what are some critical issues you can discuss? What are some potential negative, harmful, or dangerous implications of an emerging digital media environment?

2. Choose one of the 10 factors mentioned in this chapter, go to the World Wide Web, and do further research on the topic to expand your knowledge and sophistication of the topic. If appropriate, share this new and additional information with your classmates the next time you meet. Remember that this is a field of study whose content changes quickly. Statistics change, events take a turn for the better or worse, and new players enter the field. It is important to keep up on current events. One of the best ways of doing this is through the Web. You can begin by going to traditional news sites and clicking on their "technology news" link. A good example of content that changes frequently is the number of people in the United States who currently use the Internet. What would that number be today? What are the numbers for other countries? Where did you find your data?

3. Are there other factors that have contributed to the emergence of a digital media environment? What are they?

Endnotes

1. Of course, depending on how one chooses to categorize these factors, they could exceed or be less than 10 in number. These factors were distilled from a very large number of sources, many of which are in the bibliography. The factors will first be listed and then briefly explained.

2. Much of this section borrows from the author's earlier work, "Ten Things You Should Know about Digital Media: A User-Friendly Primer," published by the Freedom Forum in April 1997. It has been heavily edited as well as updated, however.

3. Many sources, including the U.S. Department of Commerce, tracks PC penetration rates. It should be noted that some states and cities have higher penetration rates than others. Commerce department reports can be found at the agency's Web site, http://www.commerce.gov.

4. Peter Elstrom, Catherine Arnst, and Roger Crockett, "At Last, Telecomunbound," *Business Week*, July 6, 1998, p. 27.

5. The Internet 2 project has its own Web site at http://www.internet2.edu.

2

Networks and Infrastructures

Early Communication and Information Networks

Although in the Digital Age, the word *network* often conjures up images of fairly complex computer linking devices, the term also has simpler connotations. Metaphorically, a network can be described as netting or mesh. *Webster's New World Dictionary (Third College Edition)* defines a *network*, in addition to the more familiar contemporary definition, as a "system of roads, canals, veins, etc. that connect with or cross one another" or "a group, system, etc. of interconnected or cooperating invidividuals." A fuller definition of network, then, should include the notion of human agency (i.e., active human participation in achieving desired outcomes) and not just technology. This chapter explores the concept of *network* from a broad perspective, showing that the exchange of news and information over networks has occurred for thousands of years, albeit through different "conduits."

The earliest networks of human communication were literally carved into the earth's surface in the form of trails, roads, and trade routes. The Silk Road, for example, spanned 7,000 miles, connecting ancient China, Central Asia, Northern India, and the Parthian and Roman Empires—essentially linking one end of the massive continent to the other. The silk trade blossomed under the Han Dynasty (202 B.C.–220 A.D.) and continued to thrive for centuries.

Spices such as nutmeg, cinnamon, and cloves were carried in ships along maritime trade routes, with ports of call in China, the East Indies (Spice Islands), the Middle East, India, and the Mediterranean, and even extending into Genoa, Venice, and Britain.

These roads and sea routes not only permitted caravans and sailing vessels to transport *goods* from one part of the world to another but they also moved *people* from one part of the world to another, and hence communicated news and ideas (not to mention language) *about* different places in the world *to* different places in the world. Then, like now, the information was not always accurate. Frederick (1993, pp. 16–17) writes of the third-century Roman Gaius Julius Solinus, who "told of horse-footed humans with ears so long the flaps covered their entire bodies, making clothing unnecessary. One-eyed savages downed mead from cups made from their parents' skulls. These depictions of foreigners found their way onto maps until the eighteenth century." Map illustrations of "foreign" inhabitants could be equally peculiar and distorted. Whether these kinds of descriptions were exaggerations of what explorers expected to find in places unknown or were intentional fabrications to scare of competitors from journeying to potential trade markets is unknown, but these examples show that misrepresentations of "the other" go back a long way in the history of mediated communication.

Sometimes traders and seafarers settled into their new environs, establishing "trade communities." Curtin (1986, p. 2) writes that "the merchants who might have begun with a single settlement abroad tended to set up a whole series of trade settlements in alien towns. The result was an interrelated net of commercial communities forming a trade network, or trade diaspora—a term that comes from the Greek word for scattering, as in the sowing of grain."

Information may have flowed at a snail's pace, by today's standards, along these ancient pathways, but they formed the template for early communication networks that still have relevance today when analyzing contemporary communication networks. These networks not only carried oral communication content (i.e., word-of-mouth communication) but written content as well. Some of the great libraries in history relied on the acquisition and exchange of books and other documents over these networks.

The Assyrian King Assurbanipal (ca. 668–627 B.C.), for example, built one of the greatest libraries in the ancient world and housed it in his palace in the royal capital of Nineveh. This famous storehouse of knowledge contained more than 30,000 clay tablets. "Under [Assurbanipal's] personal direction," library historian Michael H. Harris writes, "agents were sent to all parts of the Assyrian kingdom, which then extended from the Persian Gulf to the Mediterranean, and even to foreign lands to collect written records of all kinds and on all subjects" (Harris, 1984, p. 16).

Another example is the Great Library of Alexandria, which is said to have housed hundreds of thousands of manuscripts acquired from many parts of the world and, in a sense, was the "hub" of a regional information network in its day. (Not surprisingly, Alexandria was also a hub for trade.) Founded by Ptolemy I, a general in Alexander the Great's army, this legendary library survived for six centuries, contained the greatest collection of books in the ancient world, and attracted scholars from near and far to learn from the vast reserves of knowledge preserved within its walls.[1] It was the goal of Ptolemy I (also known as Ptolemy Soter, or the "Preserver") to bring together all the knowledge of the world into one place beginning in 300 B.C. The Ptolemies, rulers of Egypt, thereafter maintained this library from generation to generation until fire and hostile invasion led to the library's destruction in centuries later. Because this unprecedented storehouse of knowledge was centralized and attached to the palace grounds, it was close to the target of enemy forces. When the library was ultimately destroyed, the loss was incalculable and devastating. Much of the information preserved from and about antiquity was lost. As a metaphor for a highly centralized communication network, the Alexandrian library is a good example of such a network's vulnerability.

Communication networks and the emergence of information hubs, usually corresponding to centers of trade and commerce or centers of political power, continued to evolve over the centuries and encompassed a wide array of media beyond scrolls and books. Indeed, all mass media—books, newspapers, magazines, radio, and television—could be classified as network-building tools in that they link people together through a network of news and information, and certainly entertainment in more recent times (see Chapter 3). But it wasn't until the early 1800s that a telecommunications network using electricity to send encoded messages rapidly over long distances became practicable in the United States and Europe. The electric telegraph was the beginning of a modern telecommunications network and had a dramatic impact on the news industry, domestic and international business, and politics. The electric telegraph also provided the first "layer" of webbing for what would later become a global telecommunications network encompassing the telephone, radio, television, satellite communication, and, more recently, the Internet and World Wide Web.

Electricity and the Rise of Modern Telecommunications

The United States of a century ago was, like now, in the throes of a communications revolution. Riding on the momentous inventions of the electric telegraph and telephone in 1837 and 1876, respectively, wireless telegraphy was emerging as the next great wave of instantaneous communication around the turn of the century. The young Italian, Guglielmo Marconi, was the Bill Gates of his time. The son of well-to-do parents in Bologna, far north of Rome, Marconi whittled his hours away experimenting with electromagnetic radiation in the garden of his family's country estate. When his own government failed to recognize the significance of his early achievements in sending simple telegraph signals through the air (look Ma, no wires!), his mother, Annie Jameson, herself from a prominent Anglo-Irish family, arranged for a meeting with Sir William Preece, head of technological innovations for the British Ministry of Posts, which controlled telegraph and telephone communications in the United Kingdom at the time. Preece immediately saw the value of Marconi's work for the burgeoning empire: Colonies had to be administered, naval forces had to be commandeered, and transnational commerce had to be facilitated.

The British government agreed to support Marconi's work in wireless telegraphy, and eventually Marconi set up his own company—Wireless Signal Telegraph Co., later to be renamed Marconi Wireless Telegraph Co. Thus, Marconi contributed a vital piece to the emergence of national information infrastructures in the Western world that consisted of a vast and growing network of telegraph cables, telephone lines, and radio stations first within the boundaries of individual nation-states, but soon connecting those nation-states into a "modern" telecommunications network that spanned thousands of miles over vast regions of the globe. By the turn of the century, telegraph cables—some of them undersea—linked much of Europe and connected England with Canada, the United States, India, and Russia. India was linked to Australia, Cuba to Jamaica, Canada to Bermuda, and Singapore to Saigon and Hong Kong. In 1898, Marconi gave successful demonstrations of his wireless telegraph in England, Canada, and the United States, reporting on sporting events, aiding in some maritime rescues, and even linking Great Britain's Queen Victoria from her home on the Isle of Wight to her husband, Prince Edward, who was at sea on the royal yacht recovering from an appendectomy. Marconi was a master showman—he knew how to peddle his wares and reap maximum media exposure for his new invention—and soon the world was paying attention to the peculiar communication device that seemed to accomplish the impossible.

The thought of sound, music, and eventually the human voice traveling through the air from a transmitter somewhere far off in the distance, perhaps even in another country, right to somebody's radio receiver in the home must have struck some stunned observers as a kind of scientific magic. Indeed, even commercial electrical power had not been widely available for very long at the turn of the twentieth century. Thomas Alva Edison had invented the incandescent lamp in 1879, and in the years that followed, he gradually built up a system for the commercial distribution of electricity involving small generating stations. (Investors later provided the capital needed to expand this burgeoning industry.) Of course by the time, electrical current was already being used to send Morse code through telegraph wires in the United States and abroad, but as America headed into the twentieth century, the progressive series of discoveries and inventions based on the principles of electromagnetism was opening up a whole new horizon of communication possibilities. Like many modern-day technologies, such as satellites and the Internet, people may not have understood exactly how they worked at first, if ever, but the impact these technologies had on society was revolutionary on many different levels. They allowed ordinary people to communicate in ways and over distances and in speeds like never before, and they facilitated the parallel development of transportation technology, such as the railroad system and aerospace industry, not to mention the military and business sectors that were becoming omnipresent in many parts of the western world. In the United States, marked advances in modern communications technology around the turn of the century and slightly thereafter immediately preceded, followed close at hand, or accompanied other interrelated social phenomena, such as industrialization, urbanization, significant population growth (and shift), immigration, and—perhaps the most jolting barometer of social change—the first World War.

Today, we stand at the cusp of change no more bewildering than our forebears experienced but a century ago. The hype and discourse, praise and criticism, promises and perils surrounding the Internet and World Wide Web, interactive and digital television, low-earth orbit satellites, virtual reality, and distance education should sound vaguely familiar, for these narratives recycle themselves through history whenever potentially transformative new technologies emerge. The early iterations of the telephone, initially dismissed as a toy, became a serious contender to the telegraph, in the same way that many new technologies threaten to displace (but rarely do) existing ones. Critics of the telephone, however, even after Alexander Graham Bell demonstrated its functionality, assailed its lack of providing a paper record of the communications transaction, as the telegraph did. Contemporary concerns about electronic messages or e-mail not leaving a paper trail (as does traditional correspondence) hearken back to the telegraph/telephone divide. Telegraphs also required professional communicators to transmit the Morse code at the sending end and translate it on the receiving end. The telephone was so much easier to operate and the service would be available directly to ordinary people's homes. What would become of the telegraph, the telegraph operators, the message delivery boys, and the industry that had arisen around this hitherto cutting-edge workforce? The telegraph/telephone divide, then, was also about the potential obsolescence of one set of dominant communication skills and even about the justifiable concerns over workplace security and market share.

In addition to cyclical fears and concerns about the impact of new technologies, parallel technological challenges between now and 100 years ago abound. For example, the need for increased bandwidth and an ordered communications "system" was apparent from the very start of modern telecommunications. Not long after the telegraph was commercially viable were researchers such as Thomas Edison, Alexander Graham Bell, and others searching for ways to improve on the original design so that more messages could be sent over a single wire, increasing the flow of communications traffic. Edison succeeded in inventing a "quadruplexer," which allowed four separate messages to be sent over one line at one time, but Bell's work at innovation went much further: to the telephone. Yet, the telephone itself was eventually plagued by bandwidth problems, so to speak. The amount of voice traffic over telephone lines became so overwhelming that automatic switchboards had to replace women operators at the local telephone exchanges. Once the telephone "caught on" in the United States, it spread like wildfire (or maybe even like the Internet?). At the turn of the century, there were less than a million telephones in the United States; by 1924, there were 15 million. All this telephone traffic required an elaborate network of switching systems, record-keeping and billing procedures, and focused research and development to drive innovation and improvement.

These early communication technologies—the telegraph, telephone, wireless telegraph, and radio—also gave rise to the era of large communication corporations: Western Union, Bell Telephone Co., AT&T, American Marconi, and Radio Corporation of America (RCA), which later bought American Marconi and eventually became the NBC radio and then television networks. Even General Electric and Westinghouse were already major players back then, having owned significant interests at the inception of RCA.

Like all wars, World War I was no different in that it spurred development of communications technology, particularly radio, and these technologies also suffered enemy attacks, such as cut telegraph lines and jammed radio signals. Of course, many other communications technologies have benefited from their importance to military strategy, including the Internet, satellites, and virtual reality. Wireless telegraphy and later radio were invaluable for ship-to-shore and ship-to-ship communications, which otherwise had to rely on visual or aural signals. Even the *Titanic* disaster in 1912 gave a boost to wireless telegraphy when the United States passed regulations requiring passenger ships that used its ports to have wireless equipment on board, monitored at all times. Auxiliary power needed to be provided when the ships' engines were turned off.

New media technologies—such as the Internet, World Wide Web, interactive television (ITV), satellite services, virtual reality (VR), and other electronic linking devices—are laying the foundation for the emergence of an ever-expanding "communicative space."But this foundation is built on scaffolding erected over a century ago, when the earliest modern information infrastructures crossed land and sea. The communications environment today is, in many ways, like a mirror reflection of 100 years ago. Communications companies competed with each other, took each other to court, as Bell Telephone Co. did with Western Union over a patent infringement matter. The early "fathers" of commercial radio in the United States—inventors Lee de Forest, Edward Howard Armstrong, and the Canadian Reginald Fessenden, as well as the astute businessman and head of RCA, David Sarnoff—were constantly warring with each other either in court or in the court of public opinion. The radio wars of yesteryear make the browser wars of today seem almost tame. Even concerns about "universal service" are not unique to the present period but reflect earlier concerns about the provision of mail delivery, electricity, and telephone service, to rural and traditionally underserved areas of the country. And the evolution of radio from a populist technology—used by enthusiastic amateurs with an ear for innovation—to a commercial mass medium reflects trends we have seen occurring with the Internet. The corporate sector has encroached both on radio and the Internet, much to the chagrin of amateur, noncommercial users. Similar, too, is the research and development boost each of these technologies received from the federal government because of their defense-related communication functions.

Perhaps the most important parallel between now and a century ago when it comes to the transnational communications environment is the question of regulation and international cooperation. In 1865, the International Telegraph Union was formed to harmonize technical standards, set policies, and facilitate tariff agreements among European countries that networked their telegraph lines. Reflecting the growth and development of telecommunications technologies over time, the telegraph union became the International Telecommunications Union (ITU) in the 1920s and was incorporated as a specialized agency of the United Nations after World War II. This intergovernmental organization is today responsible for a wide range of international telecommunications issues ranging from the allocation of radio frequencies to discussing the equitable distribution of satellite "parking spaces" (i.e., orbital positions) in outer space. It and many other international organizations are critical entities for facilitating the development of digital networks. Others may be needed in the future. Relatively new organizations, such as ISOC (the Internet Society) and ICANN (the Internet Corporation for Assigned Names and Numbers), are helping coordinate development of Internet technologies. The ISOC's mission statement is simple: "To assure the open development, evolution and use of the Internet for the benefit of all people throughout the world."

Globally networked communications systems are a thing of the future, but they are also a thing of the past. Great minds and inventive, innovative, and entrepreneurial spirits throughout history and in the present have helped this remarkable network expand—promising great benefits to global society but also posing tremendous problems and questions at the international level. For this reason, the private sector that has laudably driven much of the world's communications development must continue to strike a productive balance against the safeguards of national communication laws (e.g., federal regulation) and international cooperation through intergovernmental organizations (e.g., the United Nations), without being unduly restrained by them. Throughout history, communications technologies may have changed, but the goal of achieving this balance between national development aims and international harmony has remained constant for more than a century and should continue to guide future evolutionary stages of the global telecommunications environment.

Telecommunication Networks Today

Telecommunication networks today are far more complex and far-reaching than ever because of the multiplicity of technologies that are now interoperable thanks to digitization and other factors discussed in Chapter 1. In addition to the emergence of a global telecommunications network, as described in the previous section, two significant national initiatives in the United States have contributed to the robust growth in telecommunications networking today. The first occurred soon after the Clinton-Gore administration officially took office in 1993. The "National Information Infrastructure: Agenda for Action" report that was released in September of that year was a call to action to create what the administration referred to as a "seamless web of communications networks, computers, databases, and consumer electronics that will put vast amounts of information at users' fingertips" ("National Information Infrastructure: Agenda for Action," 1993, Executive Summary). "Development of the NII," the report began, "can help unleash an information revolution that will change forever the way people live, work, and interact with each other."

As enumerated in the report, the NII would facilitate the following conditions in society:

- People could live almost anywhere they wanted, without foregoing opportunities for useful and fulfilling employment, by "telecommuting" to their offices through an electronic highway
- The best schools, teachers, and courses would be available to all students, without regard to geography, distance, resources, or disability
- Services that improve America's health care system and respond to other important social needs could be available on-line, without waiting in line, when and where you needed them (1993, p. 1)

The report acknowledged that the private sector would be chiefly responsible for deploying the NII—and, in fact, was already in the process of deployment at that time—but that there would also be an important role for government to play in the process. Primarily, the government would ensure that telecommunication services were available to all Americans and try to extend the long-standing principles of "universal service" to technologies such as the Internet.[2] But the report was also a rallying call. It stated unabashedly that "the time for action is now" and made no secret of its dual support for citizen access to the benefits of the NII coupled with a strong desire to support the private sector in developing the NII so that the United States would benefit from a strong telecommunications industry:

> An advanced information infrastructure will enable U.S. firms to compete and win in the global economy, generating good jobs for the American people and economic growth for the nation. As importantly, the NII can transform the lives of the American people—ameliorating the constraints of geography, disability, and economic status—giving all Americans a fair opportunity to go as far as their talents and ambitions will take them. ("National Information Infrastructure: Agenda for Action," 1993, Executive Summary)

The body of the NII report was essentially a vision statement for transformation of the telecommunications environment in the United States. It promised vast improvements and innovations in education, culture, health care, employment, entertainment, government services, and other forms of information seeking. It aimed to serve and benefit a wide segment of the U. S. population from big business to the ordinary, run-of-the-mill consumer.

The NII Agenda for Action provided an important and, in many ways, prescient blueprint for how the Clinton-Gore administration intended to build on the existing telecommunications network in the United States and take it into the twenty-first century. In later years, the NII would also be referred to as the Global Information Infrastructure (GII), suggesting the administration's vision for a global communications network. That vision is now being realized, but as the earlier part of this chapter showed, such a network has been in the making for more than a century and a half.

Another national initiative that has contributed to the growth of global networking is the Telecommunications Act of 1996, an act "to promote competition and reduce regulation in order to secure lower prices and higher quality services for American telecommunications consumers and encourage the rapid deployment of new telecommunications technologies" (Telecommunications Act of 1996, S. 652). This act sought, among other things, to lift restrictions to competition within the U. S. telecommunications sector and address the rapid changes occurring within that sector. The act was concerned with topics such as interconnectivity (which directly relates to the goals of the NII), universal service, access to the NII by persons with disabilities, and the elimination of barriers to enter the telecommunications market. It revised rules and regulations pertaining to cable television, local telephone service providers, and long-distance telephone service providers. It replaced the omnibus and increasingly obsolete and obstructionist Communications Act of 1934 and attempted to create laws more relevant to the digital media environment of the 1990s and what was to follow in the twenty-first century.

The Telecommunications Act of 1996 was not passed without considerable controversy. First, there was the underlying agenda of deregulation, which could be a benefit or a detriment to affected parties (e.g., local phone service providers, long-distance phone service providers, cable television broadcasters, traditional mass media organizations, satellite service providers, etc.), depending on what parallel safeguards were put into place to ensure that a level playing field existed among industry competitors before complete open competition went into effect. One of the fears that public interest groups had was that in a deregulated environment, telecommunications companies would just grow larger and larger through mergers and acquisitions, potentially eliminating the diversity of consumer options that the Telecommunications Act was supposed to enable in the first place. Public interest groups were not the only ones concerned. Small- and medium-sized media and telecommunication businesses worried that they would not be able to compete effectively in a market where they had not enjoyed quasi-monopoly status as had some of their competitors that had been in the business longer. This is why the act was passed only after years of debate and discussion, some quite contentious and litigious, among affected parties. To ensure a de facto remonopolization of certain telecommunications sectors does not occur, the Department of Justice and the Federal Communications Commission must both approve major mergers, as has already happened in a number of high-profile cases, such as the AOL and Time Warner merger. When a proposed merger is announced by two telecommunications or media giants, the actual merger is not instantaneous. Weeks and months may pass before the necessary approvals are granted.

Carefully worded parts of the Telecommunications Act also insist on corporate responsibility for serving the poor and disenfranchised. The universal service clause was an important provision of the act. According to the Federal Communications Commission, the goals of universal service are

> to promote the availability of quality services at just, reasonable, and affordable rates; increase access to advanced telecommunications services throughout the Nation; advance the availability of such services to all consumers, including those in low income, rural, insular, and high cost areas at rates that are reasonably comparable to those charged in urban areas. In addition, the 1996 Act states that all providers of telecommunications services should contribute to Federal universal service in some equitable and nondiscriminatory manner; there should be specific, predictable, and sufficient Federal and State mechanisms to preserve and advance universal service; all schools, classrooms, health care providers, and libraries should, generally, have access to advanced telecommunications services; and finally, that the Federal-State Joint Board and the Commission should determine those other principles that, consistent with the 1996 Act, are necessary to protect the public interest.[3]

The universal service language in the Telecommunications Act of 1996 was especially important as public interest groups worried about a phenomenon called the *digital divide*, in which segments of the population polarize toward technological "haves" and "have nots." Rural residents, the poor, underserved ethnic groups, the disabled, and other disenfranchised people could see their access to information resources eroded if content developers and service providers did not make a deliberate effort to reach out to a diverse end-user population. Although lack of money is certainly an obstacle to fully participating in a global information infrastructure, it is only one of many factors that could contribute to a digital divide. Lack of knowledge or training is another reason. What technology does a person need to get access to the Internet? How does the person set it up? How does it work? What is it good for? Educational institutions and nonprofit organizations can help bridge the digital divide by providing knowledge and training to students and community members. See http://www.ntia.doc.gov/ntiahome/digitaldivide/ for more information.

On the other end of the access spectrum has been the move to *restrict* access to certain kinds of content on the Internet. The controversial Communications Decency Act (CDA), which was part of the Telecommunications Act of 1996 when the act was signed by President Clinton on February 8, 1996, was subsequently overturned by the U.S. Supreme Court for being too broad in its proposed restrictions. Other proposals designed to restrict access to—or to create penalties for posting—certain kinds of content have also been hotly debated in political circles. More examples are discussed in Chapter 7.

The Telecommunications Act of 1996, along with the NII: Agenda for Action in 1993, set the United States on a trajectory of national and global electronic network development. It built on and advanced the scaffolding of network development that had been laid since the days of telegraph deployment in the United States and abroad, and challenged telecommunication carriers "to interconnect directly or indirectly with the facilities and equipment of other carriers" and to avoid feature, functions, or capabilities that do not facilitate interoperability.

Encouraging the private sector to work closely with government and civil society, both federal initiatives discussed in this chapter helped advance the deployment of a "seamless web of communications networks." The next chapter discusses the impact that digital media and electronic networks are having on the traditional mass media.

Questions for Discussion and Comprehension

1. Describe the evolution of communication networks throughout history. What do the ancient trade networks have in common with today's global computer networks? What are networks used for (whether ancient or modern)? What do you think networks of the future will look like?

2. There is a growing body of law alternately referred to as "Internet law" or "cyber law" or "netlaw." Research this area of law on the Web and come up with a list of different laws that exist that relate specifically to the Internet or the Web. Provide the rationale for each of these laws. In other words, what was the problem the law was addressing or trying to preempt? Why is the process of trying to regulate the Internet often difficult or controversial?

3. Search for National Information Infrastructure Agenda for Action on the Web. Read the fine details of this document and decide whether the goals of this agenda are being fulfilled. What else can be done to develop a robust and yet equitable national information infrastructure?

Endnotes

1. For an excellent history of libraries as an institution, see Harris (1984).

2. The concept of universal service goes back to at least 1934, when the Communications Act of 1934, as amended, stated that: "For the purpose of regulating interstate and foreign commerce in communication by wire and radio so as to make available, so far as possible, to all the people of the United States a rapid, efficient, Nation-wide, and world-wide wire and radio communication service with adequate facilities at reasonable charges, for the purpose of the national defense, for the purpose of promoting safety of life and property through the use of wire and radio communication, and for the purpose of securing a more effective execution of this policy by centralizing authority heretofor granted by law to several agencies and by granting additional authority with respect to interstate and foreign commerce in wire and radio communication, there is hereby created a commission to be known as 'Federal Communications Commission,' which shall be constituted as hereinafter provided, and which shall execute and enforce the provisions of this Act." [47 U.S.C 151]

3. "What Is Universal Service?" Federal Communications Commission Web site, http://www.fcc.gov/ccb/universal_service/ welcome.html.

3

Changes Affecting Traditional Mass Media

Digitizing the Traditional Mass Media

Not many years ago, a conventional mass communications course would cover media that are familiar to most of us: books, newspapers, magazines, radio, and television. Each of these media is probably in your homes; you're also holding one as you read this. Every one of these media today, however, has been affected in some way by the "revolution" in digital technologies, especially the Internet and the World Wide Web. At this writing, there are approximately 3,000 U.S. newspapers currently on the Internet. In addition, many popular print magazines, such as *Time*, *Newsweek*, *TV Guide*, *Playboy*, *Playgirl*, *Ebony*, and many others have online components. Radio stations and even individual disc jockeys have their own Web sites. And all national network television stations (ABC, CBS, NBC, PBS, Fox), many of their local affiliates, and a wide range of cable channels also have a strong presence on the Internet.

During the 1990s, many of the traditional mass media organizations realized that they needed to jump on the digital bandwagon. They often were not sure exactly *why* they needed to jump on this bandwagon, except for the fact that they did not want to be left behind even though the vehicle might well have been a runaway train.

The concern over loss of advertising revenue certainly figured into the equation. Advertisers spend more than $200 billion in the United States each year, and for more than 200 years, newspapers were the top ad medium in the nation. As overall newspaper circulation continues to decline, however, and more channels become available on cable TV, television (both broadcast and cable combined) has been taking in more ad dollars —almost $60 billion in 2000—than newspapers.[1] (If broadcast TV and cable TV are analyzed separately in terms of ad revenues, newspapers still command the lion's share.) However, newspapers have always had a revenue source against which no other medium could compete: classified advertising. Obviously, radio and television wouldn't have anything comparable to classified ads, and although some magazines do have a limited classified ads section in the back, they pale in comparison to the thousands of listings in a typical daily newspaper on any given day. Everything from lost pets to long-lost lovers appear in the classified ads—used cars, apartments for rent, homes for sale, yard sales, and "looking for a friend" ads. The threat to newspapers that the Web presented was that unlike other media: The Web was an ideal medium for classified ads. In fact, it might even be a better medium because specific ads could be located using search engines, postings could be updated frequently, and users could get access to the ads over and over again for free.

People who use the Internet to find cars, friends, homes, and so forth, say analysts, could spell trouble for newspaper revenues (assuming they are not using a newspaper Web site). Depending on the newspaper, classified ads could represent approximately 40 percent of a newspaper's ad revenues. A researcher from Forrester Research, Inc., was quoted in one newspaper article as saying in 1997 that newspapers were "going to get hammered." He said in that year that newspapers would risk losing more than $4.1 billion by 2001, which would represent 10 percent of their ad revenue, as advertisers turn to the Internet or exact discounts for staying with the newspaper.[2] As online classified ad sites were expected to rise, newspaper industry experts began to sound the alarm bell. Examples of how the Web is being used for online classified ads can be found at sites such as http://www.apartmentpeople.com (classified ads for apartment hunters), http://www.usedcars.com (for researching, buying, or selling a used car), and http://www.lostpets.com (to report a lost pet, report a found pet, search for a lost pet, and search for a found pet). Real estate ads, which are a staple item in classified ads, also find a friendly alternative on the Web, where multiple photos can be shown, search engines can be offered, and other services (e.g., a mortgage calculator) can be provided. Many real estate companies as well as individual agents have their own Web sites now, and URLs are frequently posted on For Sale signs.

Although the dire prognostications about the death of classified ads are still worth heeding, newspaper classified ads are far from dead. For one thing, many newspapers, having seen the writing on the wall, now post their classified ads on their own Web sites or partner with a vendor to post their ads. Also, newspapers continue to be the preferred medium to use to look for goods and services traditionally advertised there. But because classified ads are such an important part of a newspaper's revenues, newspapers do not want to lose the classified ad market to nonnewspaper competitors.

To maintain readership and hang on to ad dollars, many newspapers began to realize they would not be able to ignore the Internet. In the 1990s, one after another newspaper organizations began creating "new media" divisions or brought "new media" specialists on board to help with the transition to a digital media format. Magazines, as well as television and radio stations, jumped on the online bandwagon.

One can visit a number of Web sites for links to online media organizations. Three that are popular, comprehensive, and reliable are Editor and Publisher's online service, http://www.mediainfo.com (click on the "Media Links" hypertext); the American Journalism Review's AJR Newslink, ajr.newslink.org; and the Ultimate TV Web site, http://www.ultimatetelevision.com for links specific to television, and http://www.radio-locator.com for radio stations on the Web.

Traditional Mass Media and Digital Media

This chapter will look at different ways that digitization and the digital media environment have had an impact on traditional mass media. Before looking at specific media, however, Table 3.1 provides an overview of general comparisons between traditional media and digital media. These are generalities meant to draw distinctions between broad categories of traditional media and digital media characteristics. Although exceptions to these generalities exist, these distinctions are meant to be used as conceptual tools to aid in distinguishing so-called old media from new media. The characteristics listed in the left column represent traditional media; those in the right column represent digital media.

TABLE 3.1 *General Comparison of Traditional Mass Media and New Media*

Traditional Mass Media	*Digital Media*
Geographically Constrained: Content is geared to geographical markets or regional audience share; market specific.	**Distance Insensitive:** Content can be geared more toward the needs, wants, and interests of the reader, regardless of physical location; can be topic specific.
Hierarchical: News and information pass through a vertical hierarchy of gatekeeping and successive editing.	**Flattened:** News and information have the potential to spread horizontally, from nonprofessionals to other nonprofessionals, although professional online news organizations still reflect traditional media practices.
Unidirectional: Dissemination of news and information is generally one way, with restricted feedback mechanisms.	**Interactive:** Feedback is immediate and in many cases uncensored or modified; potential for more discussions and debate (or flame wars!) rather than editorials and opinions.
Space/Time Constrained: Newspapers are limited by space ("newshole"); radio and TV by time.	**Less Space/Time Constrained:** Information is stored digitally; hypertext allows large volumes of information to be "layered" one atop another.
Professional Communicators: Trained journalists, reporters, and "experts" tend to qualify as traditional media personnel.	**Amateur/Nonprofessional:** Anyone with requisite resources can publish on the Web, even amateur and nontrained communicators.
High Access Costs: The cost of starting a newspaper, radio, or TV station is prohibitive for most people.	**Low Access Costs:** By comparison, the cost of electronically publishing/broadcasting on the Internet is much more affordable.

TABLE 3.1 *Continued*

Traditional Mass Media	*Digital Media*
General Interest: Many mainstream mass media target large audiences (sometimes pejoratively referred to as the "lowest common denominator") and thus offer coverage of interest to a broad audience.	**Customized:** With fewer space/time restraints and market concerns, digital media can "narrowcast" in depth stories to personal preferences and interests.
Linearity of Content: News and information are organized in logical, linear order; news hierarchy.	**Nonlinearity of Content:** News and information are linked by hypertext; users navigate by interest and intuition rather than by logic.
Feedback: Letters to editor, phone calls; slow, effort heavy, moderated, and edited; time/space limited.	**Feedback:** Electronic mail, posting to on-line discussion groups; comparatively simply and effortless; often unedited, unmoderated.
Advertising Driven: Need to deliver big audiences to advertisers to generate high ad revenues; "mass appeal."	**Diverse Funding Sources:** While advertising is increasing, other means of support permit more diverse content; small audiences OK.
Institution Bound: Much traditional media are produced by large corporations with centralized structure.	**Decentralized:** Technology allows production and dissemination of news and information to be "grass-roots efforts" and dispersed.
Fixed Format: Content is produced, disseminated, and, depending on particular medium, relatively "fixed" in place and time.	**Flexible Format:** Content is constantly changing, updated, corrected, and revised. In addition, multimedia allows the integration of *multiple* forms of media in one service.
News Values, Journalistic Standards: Content produced and evaluated by conventional norms and ethics.	**Formative Standards:** Norms and values obscure, in formation; content produced and evaluated on its own merit and credibility.

Source: Adapted and modified from Kevin Kawamoto, "10 Things You Should Know about New Media: A User-Friendly Primer," published by The Freedom Forum Pacific Coast Center, 1997. Thanks to Professor John E. Bowes for inspiring this organizational framework.

Numerous important technical and design elements separate traditional mass media from digital media, but those differences are just the beginning. In general, digital media tend to be more multimedia (lending themselves to a convergence of technologies) and interactive, and thus the relationship between the user and the medium is vastly altered. The model that is used often to describe the changing relationship between traditional and digital media is one of "push" versus "pull" technologies. Traditional media *push* news and information (and entertainment) to mass audiences, without much regard for personal preferences in terms of what particular members of the audiences want to see, when and in what order they want to see it, and in what amounts they want to see it. Digital media are more audience driven. Content is *pulled* by the individual user at his or her own discretion. Users can access news and information based on their personal preferences and get that content when they want it, in the order that they want it, and in the amount that they want it. Often, users can also manipulate that content (e.g., save it in a file on their local computer hard drive; e-mail it to themselves, friends, or an electronic discussion group; and cut and paste parts of the information for various applications, such as writing reports or collecting research material).

Although it is theoretically correct that "anyone can be a publisher" on the Web, including amateur sleuths and news mogul wannabes, the fact is that most of the top new Web sites are not produced by amateurs but by professional, well-established news organizations such as CNN, *USA Today*, NBC (in this case, a convergence of Microsoft and NBC known as MSNBC), the *New York Times*, the *Los Angeles Times*, and so forth. All of these have comparable print or broadcast counterparts. But other news sites are principally Web based: C/Net (http://www.cnet.com), Salon (http://www.salon.com), the controversial Drudge Report (http://www.drudgereport.com), Slate (http://www.slate.com), the Freedom Forum Online (http://www.freedomforum.org), and many others. Still others, less well known, are not always considered "credible" sources of news. On the Internet, like in the real world, name credibility carries weight. *The New York Times*, whether in print form or online, has a certain degree of credibility among readers and many Internet users. But credibility is a gray area on the Internet. There was a time when few people knew about the Drudge Report, and those who did regarded it as little more than a gossip rag. Over time, the site gained more attention, in large part due to gossip reporting—most notably Matt Drudge's revelations about the Clinton-Lewinsky relationship before the mainstream news media were reporting details he posted on his site. Since then, Drudge has leaked information about a number of alleged political scandals and conspiracies, a number of which were not accurate. Nevertheless, his star has risen. He was even given a short-lived talk show on the Fox Network but parted company after ratings fell and a dispute emerged between Drudge and Fox executives over an abortion-related story. Drudge also hosted a radio talk show. His critics call him a blight in the institution of journalism and say that he has elevated gossip and rumor to the level of news. Others view him as a crusading cyberjournalist or citizen journalist who exploited the Internet to present to the public ground-breaking news and information that the mainstream news media were afraid to touch.

Although Drudge is perhaps the most sensational name associated with the nonprofessional "journalist" on the Web, it would be a distortion of reality to hold him up as representative of the ways that nonjournalists use to the Internet to disseminate news and information that don't always get covered by the mainstream media, or get covered in consistently unidimensional ways. The truth is, the Internet hosts a wide spectrum of different voices who are trying to get their message heard. One end of this spectrum could consist of the highly commercial, big-name media who now simply have their online extensions. On the other end of the spectrum could be willful propagandists or, plainly, lunatics of one kind or another. Between these two extremes exists a potpourri of other people and organizations whose funding model (or level of concern over profits), integrity, credibility, journalistic ability, and motivation vary widely. Nonprofit organizations that want to share information about certain aspects of health care, labor, the environment, human rights, and a host of other issues and concerns, for example, now have an easy and relatively inexpensive way of posting that content for a worldwide audience.

LaborNet has a Web site whose goal is to provide "global online communication for a democratic, independent labor movement" (see http://www.labornet.org). It is purported to be the "first regular Labor News web page in the United States" and is attempting to build a global communications network, an effort that is already partially accomplished through labor partners in other countries. The site looks professionally designed and contains links to forums, a news archive, labor videos, and links to other labor organizations. It looks like a news site, but, unlike the major news organizations' Web sites, there is no advertising on it. Clearly, however, it presents information from a certain perspective. From another "angle," the World Trade Organization (http://www.wto.org) has quite an extensive Web site of its own. Its point of view would come at odds with LaborNet on many issues, but each is afforded its voice and place in cyberspace. This site, too, is not supported by advertising.

After the terrorist attacks on the World Trade Center and the Pentagon, segments of American society felt that the mainstream media were not paying attention to a wide enough diversity of perspectives about how best to respond to the attacks. "Alternative" news sites, such as Alternet (http://www.alternet.org), the American Prospect (http://www.americanprospect.com), and others, offered perspectives less evident in the mainstream news sources.

CNN Interactive, Salon, Slate, the Drudge Report, LaborNet, the WTO Web site, Alternet, and the American Prospect, not to mention the thousands of Web sites and electronic discussion groups devoted to just about every topic imaginable—these are all part of the global communication network that is the Internet. They all may claim to deliver news, but today's savvy Web surfer would have to be critical of all information that is found on the Web, especially when the news is not "branded" by an established news organization. At the same time, these little-known-about information "niches" may contain information and perspectives that complement and contextualize mainstream news. The new adage that seems apt for the contemporary Web news reader is: Let the surfer beware.

Ethics and Standards

Many questions arise when talking about ethics and professional standards in digital media. Some would even argue that they don't exist! It is probably more accurate to say that they are still in formation. In the digital news media, even among those news organizations that extend their source credibility to the digital environment, there are numerous issues yet to iron out when it comes to ethics and professional standards. In a survey of online news managers of U.S. daily newspapers conducted by two journalism faculty members who were former professional (working) journalists, the researchers found that news posted online is probably less accurate than its traditional print counterpart, not only because of the speed at which online news is posted (providing less time for checking facts) but also because of staffing and training inadequacies.[3]

"Twenty-seven percent of the online daily newspapers taking part in the survey had no full-time staff members and 19 percent had just one full-time worker," according to journalism professors David Arant of the University of Memphis and Janna Quitney Anderson of Elon College in North Carolina, who surveyed online news editors in October and November 1999. Other findings: Not all online managers reported that their online print outlets follow the general ethical standards of traditional print journalism; many of them make at least some changes to material from their print edition when it goes online; and breaking news might get published online *before* it goes through traditional print editing. But the good news is that the online managers were "almost unanimous" in expecting their online journalists to have good news ethics and wanting journalism schools to require an ethics course that specifically includes online operations.

The researchers suggested that the Newspaper Association of America (NAA) adopt a "clearly marked hypertext link" that would be placed near the masthead of an online news site and contain corrections and clarifications of online news articles. They propose a "Corrections & Clarifications button" that would look the same on U.S. news sites everywhere. This recommendation arises from the finding that no uniform way exists of alerting readers to mistakes in online articles. The researchers believe that online news products must maintain rigorous standards of accuracy and integrity if they are to be credible and not damage their print newspaper's reputation.

Traditional newspapers' online products are not the only kind of online news model that exists, of course. As some of the "alternative" news models mentioned earlier suggest, digital news products come in different stripes and colors and, by design, may be less similar (and some may argue, less reliable) than traditional news products. With digital media, news and information do not have to circulate through traditional, linear, temporal, and hierarchical channels. Like the news and information that circulates at the town square or the village commons, this process may be rather circuitous and informal. In the more participatory model of digital media, where news producers are not necessarily professionalized, hierarchy and heavy editing are discouraged, advocacy and lack of neutrality may be apparent, reliability and accuracy of information may be a suspect, and credibility of the source may be questionable. With this kind of news model, digital media users need to do their own gatekeeping, editing, and verification, a consequence of the deprofessionalization of the communication relationship between sender(s) and receiver(s). This is no simple task, for it places the onus of critical evaluation on the *information seeker* rather than on established institutional conventions and norms.

Overall, digital media may provide the context and completeness of storytelling in a way that traditional media could probably never achieve. Users follow their interests and topics serendipitously or fortuitously—rather than using the linear logic that characterizes traditional storytelling—and in the process get much more (or even unintended) news and information than they originally set out looking for. Particular topics of interest can be explored to near exhaustion because of the negation of traditional space/time limitations and advertiser-driven concerns, and obscure or narrow interest topics can be made available in virtually unlimited quantity for the same reason. Production and distribution costs tend to be lower with the new media, and perhaps most importantly, a dramatic shift of consciousness may occur wherein news and information are perceived as things to be "experienced," interacted with, discussed, and argued over with others online—"pamphleteered" from every conceivable perspective. Truly the embodiment of Miltonian free press libertarianism! Such idealism has its costs, however. Concerns over privacy, quality, "truth," and accuracy have to be addressed.

It has been said that communication technologies are like a two-faced Janus: They are at once progressive and digressive, depending on how they are used. Certainly, and history provides many examples, communication technologies can be used to disseminate truth, or they can be used to disseminate lies, or even half-truths. In the digital media environment, the mechanisms for establishing truthful and ethical communication have yet to be fully formed.

The next section takes a look at how digital media impacts specific mass media such as books, newspapers, magazines, radio, and television.

From Books to eBooks

Like newspapers and magazines, books also have their online counterparts, but putting voluminous amounts of text online presents a range of problems for the book industry. Although most people would probably agree that reading the epic tome *War and Peace* on a conventional computer screen could be uncomfortable and even ergonomically unsafe over an extended period of time, there is research and developing taking place designed to make the electronic book—or eBook—more palatable to the average media consumer. A growing number of companies are already offering eBooks on the Web, although a viable marketing structure and interface have yet to be crafted for such long documents. An "On-line Books Page," hosted at the University of Pennsylvania, has links to more than 11,000 online books. An interesting experiment, this site was founded and is run by John Mark Ockerbloom, a "digital library planner and researcher," and is an attempt to provide full-length books online to readers at no charge. The works either have to be in the public domain, which means that publishing them online will not violate intellectual property (e.g., copyright) laws, or the author of the work needs to give permission for the work to be published on the site free of charge. The site's URL is http://digital.library.upenn.edu/books.

There is a long way to go before eBooks become a *commercial* success, however. Appropriate reading devices need to be developed so that reading a digital book is not a painful and cumbersome experience. For most people, the conventional desktop computer probably won't do for extended literary works. To read a long book from "cover to cover" would be uncomfortable, and the thought of spending so much time in front of a computer screen focused on one activity may not appeal to the growing number of people who already spend countless hours in front of a computer at work. Companies and research laboratories are currently trying to find out what would be the best way of digitizing a book using an interface that is both appealing to readers and commercially viable. One portable reading device, called the Rocket eBook, developed by the electronic publisher NuvoMedia, weighs 22 ounces and is about the size of a paperback book. The reader can download an electronic book into the device, store that content along with other electronic books, and then access the books at another time and place, like at the beach or on a plane. Although the business model and technologies of eBooks need to be further developed, a number of companies (e.g., NuvoMedia and the Glassbook Reader) already have working prototypes that are moving the eBook toward a more common portable digital media device. It has been said that in the near future, more people will be accessing the Internet from devices other than the conventional personal computer that sits on a desk. Portable devices, such as hand-held computers, portable eBooks, satellite-linked telephone/computer technologies, and other devices, will be the media of choice to access the Internet. When this day arrives, the eBook may become more feasible—but only time will tell.

One revenue-generating model that has been tried with eBooks has been the "pay to download" model. In March 2000, Stephen King's novella *Riding the Bullet* made history when it was released as an eBook on the Web, where potential readers could download it for a small fee (about $2.50, depending on where one downloaded it). The idea was novel, so to speak, but not foolproof. Before long, the news media were reporting that e-pirates had figured out the encrypting technology and were hacking into Web sites that carried the book and downloaded the work for free. The problem was well publicized and gave many merchants and authors alike reason to be concerned about safeguarding the author's intellectual property and securing the online book merchants' Web sites from outside attack.

In a more ideal revenue-generating model, interested readers would go to a Web site and select the titles of books they would like to buy online; they would then submit their credit card number electronically (and presumably securely) via an online form on the site; and when all was cleared and in good order, they would be able to download their book right to their personal computers. They could read the book on the computer monitor, or they could print the entire book or portions of it out (although this would almost seem to defeat the purpose of an eBook).

The model may be workable. With Stephen King's novella, the demand was so great that the Barnes & Noble's and Amazon.com's Web sites were overwhelmed. The online novella was made available on March 14, 2000, at 12:01 A.M. only on the Internet. It was written after the author's near-fatal accident the previous summer when he was hit by a van while walking near his home in Maine. An executive with Barnes & Noble was quoted by CNN Interactive as saying that "King's decision to publish his new short story in electronic format is a concrete declaration that the eBook format has arrived. We see a time in the not too distant future when virtually every book in print will be available in both physical and electronic formats." However, King was a little more circumspect: "While I think that the Internet and various computer applications for stories have great promise, I don't think anything will replace the printed word and the bound book."[4]

Traditionalists balk at the idea that the book will ever go out of fashion, and they are probably right. History is full of examples of unfounded fears that a new technology would completely replace another. In the past, there have been concerns that FM radio might supplant AM radio, that television might replace film and radio, that the compact disc (CD) might get rid of vinyl albums. Well, the last one is pretty much true, but generally the old medium that is threatened with extinction tends to find a new niche market to serve. For the most part, AM radio gave up its Top-40, classical music, and other formats, and adopted "talk radio" and news as its primary content base. Radio focused on serving local markets as NBC, ABC and CBS took over national markets. And the movie industry, although affected by the competition of television during and after the 1950s, is still going strong—but more reliant on global audiences for its profits. Clearly, film and television have a symbiotic relationship today, as films often end up as videos and television shows eventually, and television helps advertise films either directly (through advertising) or through movie reviews on the news or celebrity appearances on talk shows.

The phenomenon of eBooks may give traditional books a run for their money, but there will probably be room for both to coexist, and you can bet that traditional book sellers such as Barnes & Noble, Borders, and others are watching the development of eBooks with a cautious, if not hopeful, eye. Many problems have to be ironed out before eBooks become commercially viable (e.g., security issues, reader comfort with the user interface, author/publisher articulation agreements, etc.), but these hurdles and others are probably not insurmountable. A more interesting development, however, may be the noncommercial applications of eBooks—writers using the Internet and Web not to make profits but to disseminate their knowledge and ideas for free or low cost. The technology and mechanisms exist now to make free eBooks available to the public and, like the Online Books Page mentioned earlier, are already up and running. To what extent either the commercial or noncommercial growth of eBooks will continue is as yet uncertain, but Microsoft is banking on its success and has tried to position itself as a dominant player in the distribution of digital books. In late August 2000, it announced a partnership with Amazon.com to construct a customized "reader" through which consumers could download digital books for a fee. It had already formed similar partnerships with Barnes & Noble, Viacom's Simon and Schuster (a book publisher), and Time Warner.

Newspapers and Magazines

The print media have a longer history than eBooks on the Internet. Newspapers and magazines cautiously entered the online market at first, but when one after another started tip-toeing into the digital media pool, there seemed to be more of a collective plunge than a slow wading into that unfamiliar territory.

According to online newspaper consultant Steve Outing, who has had a column on new media news and analysis on the *Editor & Publisher* Web site since August 1995, there were about "20 newspaper online services worldwide—mostly BBSs with a handful of publisher alliances with the commercial online services."[5] (A BBS is a bulletin board system, more popular when the Internet was largely a text-based medium. The user dials up the system using a modem attached to his or her computer and could both read and leave messages on the electronic bulletin board.) In the early days of online newspapers, Outing reports, many newspapers cut deals with commercial online services, such as

Prodigy, to help launch online newspaper editions. Newspapers whose online edition started this way include the *Atlanta Constitution-Journal* and the *Los Angeles Times.* Although today the vast majority of online newspapers are on the Web, in the early years newspapers were not sure how best to make their content available on line: Through BBSs? Through commercial online services like Prodigy and AOL? Through the Web?

An example of an online newspaper service struggling to find its niche in the digital world was something called *Digital Ink.* This hybrid newspaper/digital media offspring took the same content that filled the pages of the *Washington Post* each day and "re-purposed" it for electronic distribution. Aptly named, *Digital Ink*—although not the first online newspaper—was an important experiment in media convergence. This new media content no longer produced the whiff of newsprint and hot ink during the printing process; instead, it generated a kind of buzz as billions of electronic bits of information fired through computer systems, modems, and telephone networks.

But how was the *Washington Post* content to be distributed online? Via a BBS? Via a commercial online service provider? Ultimately, the decision was made to use a service called AT&T Interchange, which required users to download software onto their personal computers and pay a monthly fee for the *Washington Post* online content. Eventually, the parties involved realized this was neither a good delivery model nor a funding model. As other newspapers went on the Web and offered their content for free, it was more difficult to justify charging for online news content. The *San Jose Mercury New's* electronic service—the now well-known Merc Center Web site—also tried a fee payment model but, like the *Washington Post,* reverted to a free service on the Web. Still, newspapers, being a business as well as a public service organ, wanted to know how their digital media offspring were going to kick back profits to the parent company. It's a question that's still being studied. Some models included:

- Monthly fee for service
- Free general service, but fee for more in-depth news and information
- Free service, but registration required
- Free service, but fee for downloading archival material
- Free service entirely, with revenue-generating advertising

Some of these models are still being used. For example, the *New York Times* on the Web (http://www.nytimes.com) requires users either to sign in or to register before gaining full access to the site. The *Los Angeles Times* allows registered site users to search the last 14 days of stories for free, but thereafter a fee is charged. The way it works is explained in the online newspaper's archives: "latimes.com Archives are offered in partnership with Qpass, our commerce provider. To display, print or download any story older than 14 days for $2.00, click on the link '*Click Here to Purchase Article.*' For additional savings, monthly subscriptions are also available." The online magazine *Slate* experimented with charging a yearly subscription fee of $19.95 in 1998 but changed its mind and offered the content for free the following year when it realized that the subscription model was not in its best interest.

The vast majority of online newspapers, however, are not making large profits. They frequently do have advertising on them, which of course is a revenue stream, but there have been no headlines screaming, "Online newspapers generate huge profits for their parent companies!" Indeed, that has been the quandary. The technological, design, and marketing hurdles have been met, to varying degrees, with success. Some of the most popular news Web sites, such as MSNBC and CNN Interactive, attract millions of visitors each month. But regardless of how many visitors a site attracts, the business end of news media organizations still want to know how these numbers and technological "bells and whistles" translate into profits.

The online magazine is another example of print technology's extension into the digital environment. Some of these online magazines are digital versions of their print counterparts—the examples here are overwhelming. Fill in just about any major magazine's name between "www." and ".com," and you're likely to find it on the Web. Some examples:

http://www.time.com
http://www.newsweek.com
http://www.forbes.com
http://www.tvguide.com
http://www.businessweek.com
http://www.glamour.com
http://www.menshealth.com

The list goes on and on. Magazines of more specific interest are plenteous on the Internet. Are you interested in cats or horses or "critters" (e.g., guinea pigs, ferrets, chinchillas, flying squirrels, and related animals you may not have even heard of)? Fancy Publications has its online service (http://www.catfancy.com) replete with articles on dozens of different animal species, photographs, discussion groups, and assorted tips on pet care. Did you enjoy playing Word Power (a game that quizzes your knowledge of words and their meanings) in *Reader's Digest* when you were a kid? Now you can go to the *Reader's Digest* Web site (http://www.readersdigest.com) and play online. There is a *National Geographic* Web site (http://www.nationalgeographic.com) that features stories, articles, and a "WildCam" that gives you a glimpse into the lives of animals through live cameras set up in a particular setting. For example, in a feature on otters, two live cameras were set up at a family's home (where there also happened to be otters living) and visitors to the Web site were invited to peer into the home via the Web cam. The explanation read: "There are two live cameras set up at the Chambers' home. The indoor camera is in the basement, where the otters eat and sleep. The outdoor camera is in the backyard, where they swim and play." The use of video on the Internet is increasing as technologies improve to make moving images more feasible. Even the ubiquitous Martha Stewart is also online with her "Living" theme Web site where one can learn how to garden, make crafts, care for pets, entertain for the holidays, and find out where Martha likes to shop. Martha also has a presence in traditional print, on TV, and on the radio (with her syndicated 90-second radio show where she tells you everything from how to make a good pie crust to why one should consider eating dandelions as a nice accent to salads). In words that only Martha could get away with, "Who knew that dandelions could be such a comforting and valuable food?"

Of course, there is a seemingly bottomless pit of pornographic content on the Internet, including the familiar magazines usually kept behind the counter at the local convenience store. Pornography, not surprisingly, is one of the areas on the Internet that is said to be profitable. Timely financial news has also shown profit potential.

The noncommercial news and information providers, on the other hand, have found a goldmine in the Internet and Web. The entry costs are minimal, compared to traditional news media, and the potential reach is global. Nonprofit organizations, educational groups, special-interest organizations, hobbyists, religious sects, fan clubs, environmental activists, alternative culture and recreational groups, and, unfortunately, even hatemongers, criminals, and antisocial movements have found in the Web an opportunity to pamphleteer, proselytize, and form online communities of one kind or another. Their use of the Web is not to earn profits but to communicate with people dispersed geographically who share a common interest. Never before in the history of humankind has the ability to communicate with so many people, so quickly, with such ease and sophistication, and in such volume been possible for such relatively low cost. Like the printing press, telegraph, telephone, and broadcast technologies such as radio and television before it, the Internet is destined to be one of history's major communications revolutions.

Radio and Television

Radio as a popular commercial medium has been around since the 1930s, and television since the 1950s. These technologies were typical of the conventional broadcast model, where a single message originates from a single source and is disseminated through the airwaves to many listeners. For decades, the traditional broadcast model dominated the ways radio and television stations communicated with their audiences.

With the Internet and Web, radio and television stations found a new way to build audience relationships. Many radio stations use the Web to better acquaint themselves with listeners. A Web page called the the MIT List of Radio Stations on the Internet, hosted on the WMBR radio station[6] Web page (wmbr.mit.edu), provides links to more than 9,000 radio stations. A search engine on the site allows you to enter your location to find out what radio stations are in your area. For example, to find out what radio stations are in the Seattle, Washington, area, you would simply enter that location in the search engine and click on the "Locate" button. The results list 54 radio stations "within close listening range of Seattle, Washington (47° 36' 23.0" N., 122° 19' 51.0" W)." Of those 54 radio stations, only 6 radio stations do not have a Web site. Station KUBE 93.3 FM is one of those that do have a Web site. On it are links to an events calendar, concert information, the station's "play list," music news, chat and video links, news about the local music scene, contests, KUBE's radio personalities and other "crew" members, information about restaurants and clubs, merchandise, and more.

There was a time when disc jockeys could remain fairly anonymous, when the only way a listner could put a life to a voice was by his or her imagination. Now, hypertext links to disc jockeys on radio Web sites give the listner a glimpse into their more personal side—what they look like and whatever other information they choose to share. One of the more personable is someone named Delilah, who invites her listeners to visit her Web site on her nightly radio program peppered with inspirational talk and an easy listening radio format. Once on her page (http://www.radiodelilah.com), the visitor is met with this message: "Hi, I'm Delilah, and I'd like to invite you to take a few minutes to browse through my Web site. Each night I do my best to ease some of the stress of your hectic day over the radio, and this Web site is a continuation of my show, and a place to share from the heart . . . Enjoy!" The site features news and information about—and even photos of—Delilah and her family, including her much talked about baby, Zack. She reads selected e-mail from her listeners on her radio show, and has seemed to successfully develop what media researchers call a para-social interaction—the idea that members of a mass communication audience can relate to media personalities on a somewhat interpersonal basis, as if they really know each other.

Television Web sites can serve a similar function as radio Web sites but have the added advantage of being used to extend a viewer's interest in a television program to the extraneous programmatic or production details that were rarely ever shared prior to the Web. Viewers of PBS, for example, have long been encouraged to visit the PBS Web site (http://www.pbs.org) to find out more about a certain subject, learn how a documentary was made, or get updates. There is an area on the site for kids, a place to make purchases of PBS-related merchandise, and places for tips on cooking, antiqueing, crafts, dog care, gardening, and more. The site also allows visitors to customize their Web browsers to local content. As soon as the user goes on the site, a pop-up window asks, "Would you like to localize your browsing experience on PBS Online so that television listings and other local features are integrated with PBS national content every time you visit PBS Online?" If so, you simply fill out an online form, and it'll do the rest.

One of the defining moments of digital media history occurred when NBC (the national and oldest broadcast television network in the United States) hooked up with the software giant Microsoft Corporation and gave birth to MSNBC (http://www.msnbc.com) in the summer of 1996. Today, MSNBC incorporates broadcast, cable, and Internet features into its 24-hour service. MSNBC is truly a prototype of multimedia convergence and conglomeration. There is a symbiotic relationship among NBC broadcast television news, MSNBC Cable (the cable television service), and MSNBC (the Internet service). They refer audience members back and forth. The same can be said for ABCNews.com (http://www.abcnews.com), the online service of ABC News. It tells visitors what's coming up on the ABC network (on its news programs or news magazines, such as *World News Tonight*, *Good Morning America*, *20/20*, *This Week*, and others), and it provides online features by ABC's celebrity news personalities. As mentioned earlier, the Cable News Network (CNN) complements its 24-hour cable service with its popular Internet site called CNN Interactive.

All of the television Web sites mentioned here (which are admittedly "big" sites sponsored by wealthy media conglomerates) are becoming increasingly multimedia and interactive. They typically feature quizzes and public opinion polls (nonscientific, of course) that help visitors gauge their personal opinions against those of other visitors to the site. For example, ABCNews.com once polled its readers about marijuana. It asked, "What policy do you support regarding marijuana use?" The answers one could choose were: legalization, decriminalization, or prosecution. Once you voted for one and submitted the vote, you would be taken to another page that showed you how others voted on the question as well as to a news story on the issue. It is an interesting way of engaging visitors to the site and to the topics of current programming or news events and controversies.

MSNBC polls its visitors about whether they would recommend particular articles to others. At the end of an article, it will ask, "Would you recommend this story to other viewers?" One can click on any number from 1 to 7, 1 being "not at all" and 7 being "highly." You can click on a link that let's you see the top 10 stories of the day, with this explanation at the top of the page: "MSNBC.com's most highly recommended stories, listed below, are rated on a scale of 1 to 7 by MSNBC.com viewers. (Results are automatically updated every 60 seconds.)" Again, this is a nonscientific method, but it engages the reader and creates a sense of interactivity—of active participation—as opposed to the more passive reception afforded to traditional broadcast media.

In addition to the interactive quizzes and polls, television Web sites often have audio and video files so that users can listen to, say, press conferences or a frantic 9-1-1 phone call or watch a car chase or, at the other end of the excitement spectrum, a droning political speech in its entirety. These features permit the news seeker to get more than just a "soundbite." Many sites also feature "live chats" (where one can talk to a newsmaker or celebrity), electronic bulletin boards (where people can leave messages for each other to read on topics in the news), and stock quote windows (where the fiscal types can get ticker tape information). Some sites even make transcripts available.

Of course, the most significant development in television has been the transition from analog to digital format. At present, most television stations are broadcasting in analog signals, a wave-pattern transmission that has been the norm since the earliest days of TV. Digital television, as the name implies, transmits signals in digital code, offering a pristine quality picture and sound. Examples of digital television often stun the viewer by its clarity, stereo sound, and more rectangular (wider) screen, akin to a miniaturized theater screen. The federal government would like all television stations to be broadcasting in digital signals by 2006. The country is now in a "transition period," but even after broadcasters begin transmitting in digital (and some already have), analog televisions will be able to receive the signals for a certain time period. A converter box is also available to allow analog TVs to receive digital signals, but the government is hoping that people will buy digital TVs, which will be good for the TV manufacturing market because analog TVs have basically saturated the market. Many households have more than one TV as it is.

But digital TV is not just about better picture and sound quality. There will be more that can be done with a digital TV because more information can travel through digital channels. The FCC explains:

> With digital television, broadcasters will have the technology available to transmit a variety of data as well as presenting television programs in new ways. This means that broadcasters will be able to offer you an entire edition of a newspaper, or sports information, or computer software, or telephone directories, or stock market updates if they choose to do so. Not only will broadcasters be able to broadcast at least one high defini-

> tion TV program, they may also, if they choose to, simultaneously transmit several standard definition TV programs. Another possibility is broadcasts in multiple languages with picture and information inserts and in some cases viewers will have the opportunity to select camera angles."[7]

Part of the transformation in TV viewing may involve the presence of interactive television (ITV). Although ITV has been made available to consumers in the past, its high cost and complexity discouraged widespread adoption. Digital TV may change that. Interactive television would allow personalized TV viewing, video-on-demand, home shopping, banking, Web surfing, interactive games, and a range of other services, including customized news and information channels. There is already a basic form of ITV available with services like WebTV.

The digital news media in all its variety are becoming a significant presence on the Internet and being accessed by more and more people as use of the Internet continues to increase. In a February 2000 Gallup Poll of a random sample of U.S. residents, more than half the respondents (54 percent) said they had access to the Internet in the past month via home, work, or school, and of these, 42 percent said they used the Internet more than five hours per week, 36 percent said they used it one to five hours per week, and 21 percent said they used it less than one hour. And 62 percent said they felt surfing the Net was a better use of their time than watching television.[8] Clearly, the audience for digital media content is growing, and the attitude of users toward the medium is generally positive.

Is the Internet a threat to traditional mass media? Yes and no. In some ways, it could be a threat if it effectively competes for a limited pool of advertising revenues, which already has to be shared among newspapers, magazines, radio, and television. It could also "steal" audience share away from those other traditional media. But the Internet could be good for traditional media if they integrate this innovation into their traditional purview, as many organizations already have as evidenced in this chapter. At one time, there were those who studied the Internet phenomenon and dismissed it as a passing fad. Few would be foolish enough to do so today. The question is no longer whether the Internet will be a significant communication and information medium in our society, but when and how, and with what effect on traditional mass media. It is, without question, already a media force to contend with and will continue to demand the attention of government, academia, business, and civil society long into the future.

Questions for Discussion and Comprehension

1. What are the advantages and disadvantages to listening to radio or watching television on your personal computer? Do you do it already? If so, how do you like it? If not, what would have to happen for you to do so?

2. Generally speaking, most textbooks are quite expensive. What if you could read your textbooks on the Web for free, albeit with advertisements on the page? Would you object to this? Why or why not?

3. Try picking up an issue of a popular magazine or newspaper, read it thoroughly, and then immediately look at that magazine's or newspaper's Web site, and read through that

thoroughly (except for the archives). Do you feel as though you are reading the same material? Was there content in one that wasn't in the other? Which did you prefer and why?

Endnotes

1. See Television Bureau of Advertising Web site, http://www.tvb.org.
2. Karen Fessler and Kevin Shinkle, "Newspapers Fortify Defenses in Internet Advertising Battle," *Minneapolis Star Tribune*, October 26, 1997, p. 4-D.
3. See M. David Arant and Janna Quitney Anderson, "Online Media Ethics: A Survey of U.S. Daily Newspaper Editors," paper presented at the Association for Education in Journalism and Mass Communication (AEJMC), August 2000, Phoenix, AZ. A summary of the study was written by Kawamoto for the Freedom Forum Online, "Survey Says Online News Lack Accuracy Checks of Traditional Print Journalism," http://www.freedomforum.org/news/2000/05/2000-05-19-09.asp, and reprinted here with modifications.
4. CNN.com Book News, "Demand for King eBook Makes Download Downright Impossible," March 15, 2000, http://www.cnn.com/2000/books/news/03/15/king.ebook/index.html.
5. Steve Outing, "Stop the Presses," *Editor & Publisher* Online, August 21, 1995. *Editor & Publisher's* URL is http://www.mediainfo.com.
6. This is the Massachusetts Institute of Technology's campus radio station, 88.1 FM, in Cambridge, MA.
7. Federal Communications Commission, Mass Media Bureau, Policy and Rules Division, "Digital Television Tower Siting Fact Sheet," http://www.fcc.gov/mmb/prd/dtv/#2.
8. Gallup Poll Survey, February 20–21, 2000, cosponsored by CNN/USA TODAY/GALLUP POLL. Findings reported on http://www.gallup.com/poll/surveys/2000/topline000220/q10t27.asp.

4

Knowing the Lingo

The Net Lingo

Most professions have a certain "lingo," or specialized language, that members of that profession use with each other and don't have to define. Sometimes the words sound highly technical or scientific, such as *epinephrine* (can you even pronounce this word?), a hormone secreted by the medulla of the adrenal gland. Or sometimes the words are slang for formal words, such as *eppy* (the nickname for epinephrine). Biochemists and many in the medical profession would know this word and could probably roll it off their tongues with little effort.

Media professionals also have their lingo. A journalist might use the term, *nut graf*, which refers to the paragraph in an article that answers the who, what, where, when and why questions. This is usually but not always the first paragraph of a news article. Or they might talk about a *news hole*, which is the space left for news on a newspaper's pages after the advertising has been laid out. Getting an article out of the *morgue* means retrieving an article from the newspaper's archives. Those who know the lingo are part of the in-group; those who do not may be regarded as uninformed outsiders.

Those in the digital media profession also have their particular lingo, and this chapter will attempt to present and define some important terms that those who study the field of digital media—and may work in it someday—should know. The list of terms here is not exhaustive, but referrals to books and Web sites that can help you increase your digital media literacy will be offered for further study. Some of these terms and the discussion surrounding them may be technical, but they are important to know if one hopes to converse knowledgeably about the Digital Age and read news stories about developments in the high-tech sector. Consider this a primer. Once you have acquired a comfort level with this chapter, continue adding to your repertoire on your own.

The Basics

The foundation of digital media is the process of *digitization*—that is, the process of converting information such as text, sound, video, photographs, colors, and so forth to computer-readable binary digits. The binary digits that computers use consist of ones and zeroes. A specific string of ones and zeroes—such as 01000001—represents something that is more familiar to us—in this case, the letter *a*—once the computer decodes the string and numbers and presents it to us in our own language.

The word *bit* comes from the term *binary digit* and is the basic unit of information (either a one or a zero) in a binary or base-2 numbering system. Eight bits equal a *byte*. Hence, the eight binary digits 01000001 is also one *byte*, which is approximately equal to the letter *a* when saved digitally in its simplest form. So if you had one byte of information stored on a floppy disk, you would have a very small amount of information indeed, only about one letter.

The word *the*, however, would take up approximately three bytes of information, or 24 bits. Likewise, 500 letters would be about 500 bytes, or 4,000 bits. Roughly speaking, 1,000 bytes equals a *kilobyte* or *KB*. (More precisely, a kilobyte equals 1024 bytes.) And a *megabyte*, or *MB*, equals about 1,000,000 bytes, or 1,000 kilobytes. The reason you rarely see the notation 1,000 KB is because it is simpler to say 1 MB. Getting bits, bytes, kilobytes, megabytes, and so forth straight requires some quick mental mathematics, but the terms are useful to know if you work in the field of digital media.

One reason it is important to keep these figures straight in your head is because when information is manipulated or transmitted digitally, it is important to know how large a digital file you are dealing with. For example, a digital file of 500 bytes is relatively small—perhaps it would contain a simple set of some 500 alphanumeric characters, a few sentences. (Actually, there are many factors that can increase the number of bits in a digital file, including the formatting, font size, use of color, and so on. In our examples, we are assuming the bare essentials—simple, nonembellished text.) On the other hand, a graphic, audio, or video file could contain millions and millions of bytes. Moreover, as anyone who has ever bought a computer knows, the random access memory of a computer and its hard disk drive capacity are measured in terms of bytes.

The following list might help you better understand bits and bytes:

Bit—a digit (either one or zero) in the binary or base-2 numbering system
Byte—a set of eight bits
Kilobyte (KB)—1,024 bytes (about 1,000 bytes)
Megabyte (MB)—1,048,576 bytes (about 1,000,000 bytes)
Gigabyte (GB)—1,073,741,824 bytes (about 1,000,000,000 bytes)

If you look at a typical high-density 3.5 inch floppy disk, you will notice that its disk capacity is 1.44 MB. If you were typing a book manuscript into a word-processing file, you could probably store about 150 to 200 double-spaced pages on this disk, depending on how much formatting you used. If you try to store much more than that, you will fill up the disk and your computer will probably interrupt your save process with a message saying something like, "Disk Full" or "There is not enough room on your disk" However, if you had a Zip drive, which has about 100 or more MB of disk capacity, you could store a hundred of those manuscripts. A CD-ROM drive, on the other hand, could store upwards of 600 or 700 MB, depending on whether data or audio or both are being stored. Most new personal computers these days have hard disk drive capacities in the gigabyte range. This is because many software applications require more hard disk drive capacity than 5 or 10 years ago.

The more multimedia elements contained in a software application (text, audio, graphics, video, etc.), the more disk space will be required. Check the "System Requirements" for some of popular multimedia software, such as Macromedia Authorware, for example. Authorware, used to create Web and online learning applications, requires 40 MB of free hard-disk space (along with a lot of other requirements). Adobe Premiere, used for professional digital video editing, requires 60 MB of available hard-disk space for installation. The software application itself takes up 30 MB. For computer users who have a number of different high-end, sophisticated, software applications on their computers, one, two, or more gigabytes of hard-disk drive capacity is no longer necessarily excessive.

There are larger disk capacities than gigabytes (e.g., terabytes), but these are relatively rare at present. However, history has demonstrated that as digital media become more complex, sophisticated, and multimedia, hard-drive disk capacity and random access memory have likewise increased dramatically. Indeed, computer "power" as a whole has doubled about every 18 months, a formula widely known as Moore's Law. In 1965, Gordon Moore, cofounder of the computer chip manufacturer Intel, made the observation that the amount of information that a microchip could hold doubles every year or so, and predicted that this trend would continue. In somewhat more technical terms, he said that the transistor density of semiconductor chips (i.e., the number of transistors per square inch on these integrated circuits) would double roughly every year. This prediction has been accurate, except in recent years the doubling of computer power has slowed to every 18 months. Today, in lay terms, we think of Moore's Law like this: Computer processing power doubles roughly every 18 months. Moreover, the cost of this processing power has decreased. This basic idea—that more can be done with computers for less money as time goes on—is essential to understanding the emergence, growth, and dissemination of digital media and digital content.

Nicholas Negroponte, the digital entrepreneur and visionary and the author of the 1995 best-selling book, *Being Digital*, has elevated the bit to revolutionary status. He has written about the difference between atoms and bits, combining literal and metaphorical references to make his point. The information age that consists of newspapers, books, magazines, vinyl albums, and so forth is one that is centered on the atom. That is, information is delivered in some physical form, tangible and apparent. In the digital information age, however, information is delivered not in the form of atoms (the physical building blocks of things you can see or touch) but in the form of bits and those electronic pulses that form the basis of computer communication. "The information superhighway," he writes, "is about the global movement of weightless bits at the speed of light" (Negroponte, 1995, Chapter 1).

As we move out of the industrial and traditional information age—an age of atoms—and into the digital information age (or what Negroponte calls the "post-information age")—an age of bits—our conventional realities are transformed into something more complex, technically and conceptually. An "address," for example, of a person you are sending e-mail to provides the illusion that a physical thing is being sent to a physical place that the message was intended to travel to, but it's really all just bits and computer networks and servers supporting the façade of a virtual reality. Our conventional notions of place and space are changed, and this change is part of "being digital."

Hardware and Software

The study of digital media technologies necessarily includes many aspects, not the least of which are the broad categories of hardware and software. By now, these terms may be familiar to most people, but it's good to review them anyway to make sure they are clear. Simply stated, digital media and communication require the presence of both hardware and software. Hardware is the tangible part of the computer system—the part of the system you can touch and feel, the equipment, the physical aspects. The main part of the personal computer itself—the case in which the hard drive and mother board and CPU chip are contained—and the computer monitor is hardware. Equipment that is attached to that main part, such as printers, speakers, microphones, scanners, joy sticks, and so forth, is also hardware. These attachments are called *peripherals.* Some peripherals are *input devices.* In other words, they are used to "put" information into the computer. The keyboard, the mouse, the microphone, and the scanner are all input devices. Some peripherals are *output devices.* The computer "puts out" information to the end-user. The monitor and the speaker are output devices.

The piece of hardware that is crucial in a computer is the *CPU,* or *central processing unit.* The CPU, also called a microprocessor, is contained on a silicon chip inside of a computer on top of a "mother board" along with a lot of other chips and widgets. It has been called the "brains" of the computer, and for good reason. It is in the CPU that most of the "thinking" or calculations take place and that determine how "powerful" a computer is. The faster the speed of the computer's processing capability—measured in megahertz (MHz)—and the greater amount of information that can be transmitted within a certain period of time (bandwidth)—measured in bits per second (bps)—are the main determinants of a computer's power. Reflecting Moore's Law, mentioned earlier, CPU power (both in speed and bandwidth) has increased considerably since the early days of electronic computing. In the early 1990s, a personal computer that had a 486 microprocessor that ran at 33 MHz was considered pretty swift! Today, for the sophisticated, multimedia computer user, that same computer would be considered ancient and impractical. The Pentium III Xeon can reach speeds of 500 to 800 MHz, and the Intel Pentium 4 is on the market. This following brief explanation from the Intel Museum in Santa Clara, CA, gives you an idea of the growth in CPU power:

> In November 1971, Intel introduced the world's first commercial microprocessor, the 4004, invented by three Intel engineers. Primitive by today's standards, it contained a mere 2300 transistors and performed about 60,000 calculations in a second. Twenty-five years later, the microprocessor is the most complex mass-produced product ever, with more than 5.5 million transistors performing hundreds of millions of calculations each second.[1]

Software—sometimes referred to as *software applications* or *software programs*—is not physical, although it may be contained within a physical device such as a floppy disk,

CD-ROM, or hard drive. Software applications and programs are computer data or instructions that are stored electronically and developed by software engineers who have been trained to write these instructions in a way that a computer can understand. Software programs and applications are developed for different reasons. Arguably the most important type of computer software is the operating system software, the application that instructs the computer how to do basic tasks, such as how to deal with input data (information passed to the computer, say, through a keyboard or microphone), how to display information, how to store information, how to recognize and utilize peripherals, and so forth. Without the operating system software, the computer wouldn't know what to do and would be functionally useless as a computer. There are many different types of operating system software for personal computers, such as DOS (Disk Operating System), OS2, Windows, and Linux.

Another important category of software is end-user applications, the most common probably being word-processing software. The term *Word for Windows* suggests two types of software—a word-processing application (Word) and an operating system software (Windows). Word Perfect is another example of word-processing software. But there are many other types of end-user application software—database systems such as Access or FileMaker Pro, spreadsheet and data analysis programs such as Excel and SPSS (Statistical Program for the Social Sciences), bibliographic software such as EndNote, and of course a horde of games and personal management or entertainment applications. These end-user applications need to be able to run on the operating system software that is on one's computer, and today most popular end-user software is compatible with system software.

An important part of running software, especially the newer multimedia types, is the computer's hard-drive disk capacity and random access memory. Without sufficient amounts of these two, you probably won't be able to run the latest software applications. The *hard drive* is where the computer stores information (and where software applications find their permanent home), and the *RAM* is a temporary storage space. *RAM capacity*, sometimes referred to simply as *memory*, is important in determining how many software applications can be running at the same time, because RAM is used whenever a software application is in use. There are different types and amounts of RAM.

When personal computers were used mostly for text and extremely simple images (sometimes created with blocks of text that were made to resemble shapes), it was not necessary to have sophisticated graphic circuitry inside the computer. As computer images became more intricate and complex, incorporating many different colors and interface designs, the information that was needed to display and co-process this complexity was put into a graphics card that attached to the computer's motherboard. In simplest terms, a video card helps interpret video data for display on a computer monitor. Early standards for low-resolution and color-limited graphics were the Color Graphics Adapter (CGA) and Enhanced Graphics Adapter (EGA). The Video Graphics Array (sometimes called the Video Graphics Adapter or VGA) in the late 1980s ushered in technology that enabled higher resolution and more colors than ever before. This standard has since been improved on by video graphics developers to enable even crisper and more versatile graphic resolution. Today, a variety of advanced video card technology is available to computer users.

Look at the following list of system requirements for Adobe PageMaker Plus. You should easily be able to decipher what it is they are saying now.

- Intel Pentium or faster processor
- Microsoft Windows 95, Windows 98, or Windows NT 4.0 or later
- 16 MB of RAM available to PageMaker on Windows 95 or 98 (32 MB recommended), 32 MB of RAM available to PageMaker on Windows NT 4.0 (64 MB recommended)
- 14 MB of available hard-disk space for minimum installation (175 MB of free hard-disk space for full installation)
- CD-ROM drive
- VGA display card (24-bit or greater recommended)
- Adobe Postscript language printer recommended

Basically, the requirements demand a very fast processor (no 486 microprocessors need apply); a recent Windows or Windows NT operating system; 14 MB of hard-disk capacity; 16 MB or 32 MB of RAM, depending on what operating system is being used; a CD-ROM drive (hardware); a VGA display card (software—well, the card itself is hardware, but the instructions in it are software); and a printer (peripheral). This is the basic lingo of the personal computer system. There are other terms, of course (*ports, expansion cards, buses, plug and play, scanner, cache, input and output devices,* and various networking terms—look them up if you don't know them, using the Web site suggestions at the end of this chapter), but the basics of personal computing have been covered.

Networking and Digital Media

What makes digital media so enjoyable and useful for most people are not the technical terms just discussed, but what users can do with the hardware and software once they are up and running. The rise of digital media was reliant on advances made in hardware and software, as well as in networking technology, which is the final major piece in this chapter. (How digital media has been adopted by many in society on a more personal level is the subject of the following chapter.) Networking permits computers to communicate with other computers, people to communicate with computers and vice-versa, and the exchange of news, information, conversation, entertainment, and commerce on a local, national, and global scale.

There have been volumes written about the history and evolution of computer networking. This section provides first a broad overview of networking and then specifically looks at this technology in relation to the Internet and World Wide Web (WWW). A computer network exists when two computers are able to communicate with each other, or when 200 million computers are able to communicate with each other, and everything in between and beyond. This ability was not built into the earliest computers, which were essentially "stand-alone" computers—remarkable for their day but obviously limited by today's standards. For a comprehensive treatment of this topic, many good books exist, such as Martin Campbell-Kelly's *Computer: A History of the Infor-*

mation Machine (1996), Paul Ceruzzi's *A History of Modern Computing* (1998), and Peggy Aldrich Kidwell's *Landmarks in Digital Computing: A Smithsonian Pictorial History* (1994).

One could probably find several potential candidates for the "first" computer, beginning as far back as Blaise Pascal's Pascaline calculator in 1642 to Charles Babbage's more complex mechanical contraptions in the 1800s. These and others of their time period were mechanical computers, powered by human energy and a series of springs and gears. The first generation of computers built beginning in the late 1930s, however, laid the foundation for modern electronic computing. Physical monstrosities by today's standards, these computers (with names like the Mark I, Colossus, ENIAC, EDVAC, and UNIVAC) were important for military-related data crunching, but they also provided a kind of technological scaffolding upon which computer engineers could build and expand. Moreover, during the second half of the 1950s, sophisticated programming languages evolved and were applied to computing. Languages such as FORTRAN (1956), ALGOL (1958), and COBOL (1959) were developed, and computers gradually moved into the commercial sector for large-scale data processing. The computer was on its way to becoming a household appliance, although few if anyone would have suspected it at the time. For one thing, computers were heavy, ugly, and clunky—hardly something you'd want sitting in your living room. For another, why would anyone need a computer at home?

In the annals of computer history, a critical development in digital communication occurred with the conceptualization of computers as *networking* technologies. Electronic networks, of course, were not unfamiliar at this time. The United States had gone through various phases of telecommunications network development, beginning with the telegraph around the mid-1800s and the telephone some decades later. Commercial telephone service began on a relatively small scale in 1877. At the end of 1899, American Telephone and Telegraph (AT&T) became the parent company of the Bell system, and eventually the AT&T monopolization of the national telephone network would be set into motion.

After telegraphy and telephony, the commercial broadcast radio network arose in the 1920s after advances in "wireless telegraphy" and "wireless telephony." In the twentieth century, the traditional media of telephone, radio, and later television comprised the dominant communication networks, but they were all analog technologies and thus gave way to analog communication networks.

Analog refers to the way that a communication signal is transmitted. The signal is represented as a continuous wave pattern based on varying frequencies and amplitudes. Depicted graphically, analog transmission looks like a series of connected and fluid waves; the height and length of those waves fluctuate in accordance to the actual fluctuations of the information being transmitted (for example, a segment of music). Digital signals, as discussed earlier in this chapter, consist of ones and zeroes—discrete (or separate) bits of information, not continuous like analog. Digital transmission requires that information be first digitized before being transmitted from the sender side and then "undigitized" on the receiving end. That's the job of a modem, which stands for "modulate/demodulate."

If a human voice were recorded on an old-fashioned cassette tape recorder and mailed to someone across the country, there would be no need to "modulate/demodulate." The sound of the voice is "attached" to the cassette tape in analog form. On the other hand, if this voice were recorded into a computer and sent across the country as an e-mail attachment, it would first have to be modulated (made into a digital file of ones and zeroes), then transmitted across the country through a conduit of sufficient bandwidth, and then demodulated (to convert the ones and zeroes into a human-sounding voice again).

As computer-related research and engineering evolved after World War II, the ability to link computers to each other gave rise to another kind of network—namely, digital networks. This type of network carried bits and bytes over its conduits. Information—regardless of what form it took in the real world (text, sounds, video, graphics, etc.)—could be digitized as ones and zeroes and sent over conduits as electrical impulses at the speed of light. For this occur, however, computers needed to be able to share information with each other.

Computer networking takes much more than just hardware, of course. You cannot just cable two computers together and start them "talking" to each other. The pioneering work of computer network engineers revolved around how to develop a set of rules by which computers could communicate with each other, even if the computers were made by different manufacturers. This involved getting them either to talk the same language or to be able to translate each other's language. The set of rules that facilitate this interoperational computer networking is called a *protocol.*

The pioneers who helped develop the protocols and infrastructure that led to the existence of the Internet, a global network of computer networks, were a dispersed group of government researchers, scientists, and academics working collectively toward the creation of a robust, decentralized, complex national computer network. Beginning in the early 1960s, this group began laying the foundation for what was to become the Internet as we know it today, except back then the vision of what they were creating was by design more narrow and restricted. Against the backdrop of Cold War animosities between the Soviet Union and the United States, as well as U.S. Department of Defense initiatives to build a national communication infrastructure, the protocols that gave rise to the Internet were born.

Many names are associated with this early period of computer networking: Leonard Kleinrock, Robert E. Kahn, Vinton Cerf, and others. An enormous amount of information has already been published about the history of the Internet and will not be repeated here. (For one history written by a number of the pioneers themselves, visit the Internet Society homepage at http://www.isoc.org.) Some key events will be highlighted here, however. The D.O.D.-controlled ARPANET (Advanced Research Projects Administration Network), started in the early 1970s, was the foundation of the current-day Internet. ARPANET was designed to conform to an "open architecture" internetworking environment. Simply put, this meant that networks (for example, local computer networks) could connect to and participate in the larger ARPANET network, regardless of configurations at the local level. It was able to do this because of a protocol, which allowed the internetworking among many different networks and types of networks, called Transmission Control Protocal/Internet Protocol (TCP/IP), developed by Robert Kahn. TCP/IP replaced an earlier protocol that was less conducive to open architecture. It is because of ARPANET, TCP/IP, the philosophy of Open

Architecture, and the hard work that many people put into building this network of networks that the Internet exists today.

Around the same time that "internetting research" was going on, computer networks on a different scale were being deployed. Relatively small networks called *local area networks (LANs)*, which could connect computers in a room, office, building, or some such self-contained area, were possible thanks to network software protocols. A very popular LAN protocol developed by Xerox and others is Ethernet. A LAN could then be connected to a wider network of LANs, perhaps in a city or region. This larger network is called a *wide area network or WAN.*

Essential to understanding the difference between the telephone network and the Internet computer network of networks is a technology called *packet switching.* A telephone network uses circuit switching, which requires a dedicated channel or circuit every time a phone call is made. When Jeff in Chicago calls Sheryl in Seattle using a telephone, a dedicated line from Chicago to Seattle is established for the duration of that phone call. When the call is over, that circuit can be broken. (A *circuit* is a path over which electrical current flows.)

If Jeff sends a message to Sheryl via e-mail, however, a technology called packet switching is used instead. His message is broken up into digital "packets" that get sent through a variety of conduits—computer networks, telephone lines, even wireless transmission. When these packets arrive at their destination, they are restored to a whole message so that Sheryl can read it. As long as Sheryl gets the message as Jeff sent it, she shouldn't care how it got to her. Packet switching creates the illusion that a message arrived via a dedicated circuit, but it didn't. TCP/IP protocol made this type of communication possible.

As has been said and demonstrated many times by now, the growth of the Internet has been, literally, exponential. It would be futile to put an exact number of Internet users in a book such as this, even if such a number existed, because of the rapidly changing nature of Internet demographics and the range of methodologies used to estimate Internet usage. Even before this book went to press, the number would be outdated. Based on a review of a number of different surveys, however, it is probably safe to say that the worldwide total of Internet users consists of several hundred million people, and the majority of those users are based in North America.

There are many reasons for this growth—some technological, others social, political, and economic. It is really a convergence of factors that has led to the Internet being what it currently is. (See Chapter 1 for concise list of factors.) But certainly most analysts agree that two significant factors that have contributed to the Internet being a popular consumer medium has been the emergence of user-friendly computer interfaces and, relatedly, the development of the World Wide Web. Since the late 1970s, due to advances in technology and a reconceptualization of potential markets for computer equipment, the computer has become smaller, more powerful, and much easier to use for the nonspecialist.

Apple Computer, in particular, had the nonspecialist in mind when it created its *graphical user interface (GUI)*—the graphics that appear on the computer monitor to help a computer user operate the system and program software. Rather than force the user to use nonintuitive text commands, GUIs showed little pictures (or icons) that were linked to commands, allowed the user to "point and click," and generally made the relationship between person and computer less intimidating and even enjoyable. Windows also uses a GUI, remarkably similar to what Apple earlier used with commercial success. The 1980s also ushered in a software explosion that has yet to subside. Software for the personal user abounds, as a visit to any computer store will bear out. Indeed, many software applications can be downloaded for free or low cost right off the Internet itself.

This brings us to another interface—perhaps the mother of all interfaces—namely, the World Wide Web. People sometimes refer to the Internet and the Web as if they were the same thing. They are not. The Web is part of the Internet, which is the much larger category. But there are many parts of the Internet that are not the Web. Tim Berners-Lee, while at the European Particle Physics Laboratory (CERN), proposed a global hypertext project in 1989, based on previous work he did at CERN. In the summer of 1991, the World Wide Web—a hypertext browser/editor—was made publicly available. Thanks to his work, those who have the capability and access are able to search for, find, download, and otherwise use documents on the Internet that are coded in Hypertext Mark-Up Language (HTML) using Web browsers such as Netscape and Internet Explorer. An early popular browser was known as Mosaic, developed at the National Center for Supercomputing Applications (NCSA) at the University of Illinois at Urbana-Champaign.

Web sites have a unique "address" called a *uniform resource locator (URL).* For example, the URL for the CNN Interactive's Web site is http://www.cnn.com. The part of this address after http:// (which stands for *hypertext transfer protocol*) is CNN Interactive's *domain name*—namely, www.cnn.com. A computer on which a Web site is hosted has what is called an *IP address* or *internet protocol address,* which is a series of numbers that identifies a specific network and host, but most people today opt to use the easier-to-remember domain name to identify the "location" of their Web site. The nonprofit organization responsible for managing the domain name system is known as the *Internet Corporation for Assigned Names and Numbers (ICANN).*

Other procedures that can be done on computer networks include sending or retrieving files from one computer to another using *FTP* or *file transfer protocol.* Connecting from one computer to another computer using Telnet is also a networking function. People traveling on business or students and educators away from college campuses for holidays can often check their e-mail from distant locations by finding a computer that has Telnet on it, and then using that computer to connect to their company or school computer network server as if they were using a terminal directly connected to that server. This is why Telnet is referred to as a *terminal emulation program.* It pretends to be a terminal connected to a network server, even though it is really a remote connection.

We're Watching You

The Internet is ideal for surveillance and monitoring technologies because of the networked nature of computer-mediated communication. This is a lingering problem for privacy advocates who fear private, personal information may be too easily accessible to people who should not have that information. One of the most common monitoring devices is called a *cookie*, which gets saved in the memory of a user's browser. That's why it sometimes seems that a Web site "knows" whether someone, or an IP Address (the computer's identification to other computers) has visited it before. It is recognizing information that is stored in your browser's memory. These days, most people who go Web surfing have had cookies planted in their browser's memory whether they know it or not. It is possible to set up your browser to reject cookies. For a thorough explication of cookies, including how to get rid of them or reject them out of hand, visit the cookiecentral Web site, http://www.cookiecentral.com.

Cookies are symbolic of growing concern about privacy on the Internet. There are many other technologies that are more problematic, unfortunately. Recently, there has been a rash of computer viruses that get sent to unsuspecting people's e-mail address with enticing subject headings such as "Thought you'd enjoy this," or "Here's that information you asked for." When the recipient opens the attachment, it unleashes a program that creates havoc with the computer system, everything from erasing files to sending out messages to everyone in a person's electronic address book. It seems that more and more warnings are being issued about these kinds of viruses being spread over the Internet.

The technology exists for monitoring what people do online, for cataloging personal data, and for measuring workplace productivity (e.g., counting the number of keystrokes performed by an employee per day or per hour, etc.) Web servers keep logs of who visits the Web site, from where, and other information. Much of this collected information is used for market research, for improving services or for personalizing content. But privacy advocates warn that people shouldn't become too complacent about surveillance and monitoring technologies that purport to be beneficial to end-users. The instrumentation of Big Brother may not be video cameras, after all, but perhaps the personal computer. Personal information collected online could be misused—deliberately or by accident—by businesses, government, and other organizations in ways that might harm a person's reputation, financial status, credit rating, and right to be left alone.

Future consumers, creators, distributors, and managers of digital media content would do well to familiarize themselves the intricacies, potential, and possibilities of digital communication well beyond what is in this chapter. This has been an overview and an introduction of key terms that a student and future professional in digital media will be expected to know, but it is only a beginning. There is much more to learn. For those who choose to pursue studies or work in digital media, this gives you a start in learning the "lingo."

The following chapter will show how digital media are being integrated into many people's personal and professional lives and will contain more relevant terminology. The following is a list of terms you should know after reading this chapter. If you need more information about any particular term, use the suggested Web sites (listed after the terms) to look up the word and read more about it. You can also use those sites to look up other unfamiliar technical terms that cross your path in the course of your study, outside reading, or special projects. As the adage goes: Never let a strange word pass unchallenged! (In other words, look it up!)

Terms You Should Know

Analog
Bandwidth
Bit
Byte
Circuit switching
Cookies
CPU
Digital
Domain Name System
DOS
File transfer protocol (FTP)
Gigabyte
Graphic user interface (GUI)
Hard drive
Hardware
HTML
HTTP
ICANN
Input device
Internet
Kilobyte
LAN
Linux
Megabyte
Moore's Law
Network
Nielsen/Net ratings
Open architecture
Operating system
OS/2
Output device
Packet switching
Peripherals
Protocol
Random access memory (RAM)
Software (system and end-user)
Telnet
Transmission Control Protocol/Internet Protocol (TCP/IP)
URL
VGA
WAN
Windows
World Wide Web

Useful URLs

How Stuff Works by Marshall Brian (http://www.howstuffworks.com/bytes.htm)

Webopedia (http://www.pcwebopedia.com/)

Netlingo (http://www.netlingo.com/)

TechWeb's Tech Encyclopedia (http://www.techweb.com/encyclopedia/home)

How Stuff Works Web Site (http://www.howstuffworks.com)

Office of Learning Technologies (Human Resources Development Canada) Glossaries of Learning Technologies Terms (http://olt-bta.hrdc-drhc.gc.ca/info/glosse.html)

Questions for Discussion and Comprehension

1. How has everyday language changed since the emergence of digital media? What kinds of words are in your vernacular that are part of an Internet or Web cultural linguistic system?
2. Choose three or more of the terms that were discussed in the chapter and do further research on them on the Web.
3. If you were asked to elaborate on the history of computer technology from about World War II to the present, what would you say? (You may have to do more research on the Web.)

Endnote

1. From the author's visit to the Intel Museum.

5

Digital Media in Our Lives

Quiet Revolutions

The previous chapter provided examples of terminology that should be part of the digitally literate person's vocabulary. This chapter takes a more practical approach and looks at the world around us for examples of how digital media have penetrated our everyday lives.

A student entering college in the early 1980s would have encountered a very different college campus than exists today. For one thing, the idea of electronic mail—so ubiquitous today on college campuses in the United States and abroad—would have been unheard of among most of the campus population. Unlike today, when students routinely e-mail their friends, family members, and teachers from computers in dorms or libraries, or campus computer labs (or home computers), back in the early 1980s the technology of e-mail (and of personal computing, in general) was still relatively remote for all but the most elite computer users on campus.

Indeed, many libraries were still on the card catalog system, a seemingly archaic technology by today's standards. To find a book, students would approach a centralized hub of cabinets with drawers holding index card-sized records through which one could manually search by subject, author, or title. This was to change in the 1980s as many libraries began converting their card cataloging systems to electronic form. Barbara Moran (1984, p. 5) described this change in a book called *Academic Libraries: The Changing Knowledge Centers of Colleges and Universities.* She said that academic libraries were in the middle of what could be called a "quiet revolution":

> Today's libraries are in transition from manual to electronic systems.... [Electronic] databases are replacing card catalogs and printed indexes and abstracts. Information is being produced and stored in new forms. The merger of computers and printing is leading to a new method of information transfer. Libraries are no longer self-sufficient but are linked through electronic networks of various types. The changes brought about by

> advances in technology have been so extensive that it is difficult to assess their total effect, but it is clear that libraries are in a stage of fundamental transition.

It wasn't just the card catalog system that was in fundamental transition. Automation of bibliographic records had a domino effect. In addition to cataloging procedures, book fund accounting and acquisitions, circulation, book security monitoring, book maintenance and repair records, statistics gathering and management information, interlibrary networking, and the streamlining of internal library procedures and form controls were all being transformed and restructured by automated systems. Some libraries experimented with automation as far back as the 1960s (e.g., the Information Transfer Experiments, or INTREX, project at MIT; the Machine Readable Cataloging, or MARC, at the Library of Congress; and later the Ohio College Library Center, or OCLC, now known as the Online Computer Library Center), but these were pioneering efforts. The decision by the Library of Congress to close its traditional card catalog, begin using an automated cataloging system, and adopt new rules for cataloging beginning in the 1980s persuaded other academic libraries to follow suit.

The electronic revolution couldn't have happened at a better time for libraries, many of which were bursting at the seams trying to accommodate the ever-increasing volume of books and other holdings that needed to be housed, documented, and otherwise managed. The process of converting to electronic systems may have struck many libraries as being labor intensive and unwieldy in the short run, but it was necessary over the long haul to exploit the benefits computer technology had to offer so that information could be accessed and managed in more expedient and systemized ways as library holdings grew to sizable proportions. Of course today, a growing amount of information in libraries that used to be available in hard copy is now available online either exclusively or concomitantly through CD-ROM databases, proprietary networks, or sites on the World Wide Web.

The ability to store and manage information efficiently is an obvious necessity in a world where information accumulates rapidly. Academic publishing alone accounts for a considerable proportion of the largesse. This following comment, which appeared in the Foreword to the *Publication Manual of the American Psychological Association* (3rd edition), clearly illustrates the extent of academic publishing over the years: "The 1929 guide could gently advise authors on style because there were then only about 200 authors who published in the 4 existing APA journals. Today [this was written in 1982], the editors of APA's 18 journals consider close to 6,500 manuscript submissions a year." In another example, Moran (1984, p. 2) writes, "In 1972, for example, Cornell University noted that it had taken 70 years to acquire its first million books, 20 years for the second, nine years for the third, and six years for the fourth."

Clearly, the amount of information in the world does not decrease or remain static; it grows and, in a sense, reproduces itself as existing knowledge begets new knowledge. Like the expanding universe in the natural world, the universe of information keeps getting larger and larger. It was not possible to contain and manage this swelling beast within concrete walls indefinitely. Digitization provided a timely solution.

The library is an interesting microcosm for the way that digital media—infrastructure, content, and institutions—have been deployed throughout society as a

Online Catalogs
Most libraries have comprehensive automated systems for cataloging, circulation, and other functions. Patrons go to computer terminals rather than card catalogs to help them locate library resources.

whole. The process has been gradual but effective, a "quiet revolution." Today, the automated library is what students and faculty consider the norm; the card catalog is a remnant of yesteryear. Will the online newspaper be the norm of tomorrow? That is, as of yet, not certain. Information process models in society, however, are clearly changing. Whereas the old information paradigm saw the tools and substance of knowledge building located in a centralized, protected enclave within repositories of brick and mortar (e.g., libraries, museums, archives, temples, etc.), the *new* paradigm welcomes infiltration from outsiders through a series of digital paths and networks, unless, of course those pathways were meant to be restricted from outside "infiltration."

In a provocative article published in *Science* magazine, Columbia University professor Eli M. Noam discussed the origins of the old paradigm, linking its development to the ancient libraries of Ninevah and Alexander. He wrote, "This model—centrally stored information, scholars coming to the information and a wide range of information subjects housed under one institutional roof—was logical when information was scarce, reproduction of documents expensive and restricted, and specialization low" (1995, p. 247). It was a model, he wrote, that remained stable for more than two-and-a-half millennia. "Now, however, it is in the process of breaking down."

Digital media technologies have broken an information process model that was dominant throughout the history of humankind, beginning in an era when information was first collected in any organized and systematic manner. The consequences of this paradigm shift are many, but the chief characteristic of a digital revolution is that never before in the history of humankind has so much information been accessible by so many people with such ease at relatively low cost. In addition, never before have so many people had the ability to communicate with other people over short or long distances with such ease at relatively low cost. Advancements in technology, supportive social policy, and consistent market demand over time have contributed to the widespread success of this new paradigm.

Digital Media as the Norm?

Today, the consequences of digital media are all around us. If the Internet could be compared to a kind of global library, the resources within that library reach the user via the new information processing paradigm (i.e., through a personal desktop computer or, increasingly, other interfaces, such as a hand-held computer or an information kiosk). Those resources, seemingly infinite and boundless, are not centrally housed and restricted. Rather, they are remotely accessed and transported as digital bitstreams. Advances in digital media technologies at many different levels (e.g., networking, interface and navigational design, bandwidth, etc.) have made the trafficking of information (including news, entertainment, and even commercial transactions) a relatively "friendly" and straightforward process with the right equipment and training. People are getting access to that equipment and training from work, school, clubs, churches, neighborhood organizations, local libraries, and other places.

Many Internet users in the United States have access from home these days, but even those without home computers may access the Internet from school, work, or public access centers. According to various sources, more than half of U.S. households have personal computers, and more than half of the respondents surveyed by the Gallup Organization in February 2000 (mentioned earlier in Chapter 3) said they were current users of the Internet. Of these, 72 percent said that the Internet had "has made their lives better." (Only 2 percent said it made their lives worse, and 26 percent said it made no difference.) This well-known polling organization randomly selected its subjects for this survey, so the findings are likely to be reflective of the population as a whole.

Nielsen//NetRatings supported the findings of other research organizations. A July 2000 press release said, "Internet usage has reached critical mass in the U.S., with 52 percent of the home population having Internet access and 32 percent of the home population surfing the Web in July. Nearly 144 million people in the U.S. had access to the Internet from home, compared to 106.3 million a year ago in July 1999, a growth rate of 35 percent over the past year."[1]

The demographic of Internet users has been changing as well. Images of the computer scientist or high school geek, usually male, as the typical Internet user have long been inaccurate. In August 2000, two Internet research companies—Media Metrix and Jupiter Communications—jointly announced that "the number of women online surpassed that of men for the first time ever" and that "the population of women online is growing more rapidly than the online population overall." Research conducted by the two companies showed that "the most dramatic growth during [the past 12 months] occurred among teenage girls aged 12–17, which increased from 1,956,000 to 4,426,000 unique visitors, a percent change of +126.3%."[2]

Understandably, the number of young people who use the Internet has skyrocketed over the years, in large part because of the growth of computer facilities in schools and universities. Many schools at the K–12 level are making computer technology available to their students in some capacity, and colleges and universities are also struggling to maintain adequate technology support to students, faculty, and staff. In addition to supporting campus or departmental computer labs, colleges are wiring dorm rooms, classrooms, and faculty offices, and some programs are even *requiring* students to have access to personal computers as a condition of entry. Some universities have formed themselves into consortia, using distance education technologies for allowing students to access resources from each other's respective campuses. Indeed, distance education itself, once held at bay by mainstream universities, is being taken more seriously by respected colleges and universities as a alternative educational delivery method and revenue stream.

Research on Internet usage suggests that at some point in the near future, if not already, a large proportion of the U.S. population will be using digital media in one form or another. But the transition to this new media environment will not be accompanied by drum rolls and fireworks. Like the automation of various library systems, the transition will be an extended series of quiet revolutions, until a time when it will be difficult to imagine a period before digital media—before e-mail, before the Internet, and before Web sites. Like other technologies in the past that were at one time thought of as "new" (think of the telephone in the late 1800s, the broadcast radio in the 1920s, and the television in the 1950s), digital media eventually will be more taken for granted than regarded as exceptional. Developments now are setting the groundwork for this new media environment.

The Digital "Groundwork"

The foundation is being laid for a digital media environment that may be commonplace in the future but, for now, is still innovative enough to be studied by contemporary students of communication. We've already talked about the ways that digital technology is transforming educational institutions, which is important because such institutions are influential change agents in society. But digital technology is also transforming the home and workplace.

Today, many new homes are being constructed with advanced communications technologies in mind. So-called *smart homes* have conduits and capabilities built into them for the savvy technology mavens who insist on having an integrated network of multimedia services around them and, perhaps most important, quick access and large bandwidth. The grandmother of all smart homes is the one Bill Gates has had built overlooking Lake Washington in Seattle. Gates has been quoted as saying that in the future, "the home itself will almost be like a computer system." He should know. His $60 million waterfront estate, now almost the stuff of folklore, contains miles of computer cable hooked up to a network server that can customize the house to its occupants' desires. The *Seattle Times* described Gates's house as being filled with "smart gadgets," a place where visitors will put on an electronic pin that wirelessly alerts a computer system where the guest is in the 20,000-square-foot mansion so that it can change music, artwork, lighting, and temperature accordingly. Phone calls can be directed to the nearest phone.[3]

Are these homes the stuff of science fiction? Not anymore. You don't have to be Bill Gates to have a home filled with appliances and media controlled by central networking. In the late 1990s, the term *smart home* came to denote the kind of home that was wired for the new millennium. A large color photo in the March 7, 1998, issue of the *Seattle Times* showed a man handling a large bundle of colorful cables that he was installing in a home in Redmond, Washington (where, incidentally, Microsoft's headquarters is located). The caption read: "This fistful of multicolored wires will provide capability for computer, phone, audio, video and home-automation hookups."

The *San Diego Union-Tribune* featured an article titled, "Don't Expect Every New House to Be a Smart House, Just Yet," in which it described technology that could "centralize the computing power in a house and harness it to the comfort and conveniences of the inhabitants."[4] It gave as an example a house retrofitted by IBM and a local installation partner in San Diego. Using IBM's technology, the article said, a homeowner "can network virtually every utility, appliance, home entertainment system and computer" in the house. "Lighting, air conditioning, heating, security cameras, stereo, refrigerator, TV—virtually everything wires back to the box." You can control these myriad devices through the TV or personal computer. Condominium buildings are sprouting up with networked wiring as one of its features. Smart homes and condos are not inexpensive dwellings, and people with more modest resources may only be able to technologically "smarten" their homes on a limited basis (if at all), but perhaps someday "homes with brains" may become more commonplace, especially as entire neighborhoods and cities begin wiring their residents into virtual communities.

At one time, the idea of a "wired city" or a "wired neighborhood" was novel. Today, one easily loses count of how many there are. One of the early examples of a wired community was the Blackburg Electronic Village, a project conceptualized in 1991 and "built" by Virginia Tech (a.k.a. Virginia Polytechnic Institute), the Town of Blacksburg (in Virginia), and Bell Atlantic telephone company. This pioneering electronic village, which became a model for other communities wanting to wire up, was launched in late 1993, expanded, and continues "to foster the virtual community that has been created to complement and enhance the physical community, according to its website." It has information about Blackburg's schools, museums, libraries, government, health-care providers, and all manner of civic life. It has also hosts electronic discussion groups from the environment to business affairs. There are places on the Web site to post or get information about goods and services in the town of Blackburg.

Ideally, the residents of Blacksburg can surf the Blackburg Electronic Village site to satisfy many of the information and communication needs. For those who are not familiar with the workings and trappings of the Internet and its diverse potential, the site offers information about special classes and programs designed to initiate them into the world of digital media. By now, hundreds of villages, towns, cities, and regions have set up their own Web sites "to complement and enhance the physical community." (Take a tour of this comprehensive Web site for yourself: http://www.bev.net.)

As smart homes and electronic community networks become more prevalent, so will the technology that allows people to increase the flow of data traffic to and from their homes onto the larger network of networks. This is important because a troublesome obstacle to high-quality multimedia networking from the home has always been inadequate bandwidth and the resultant sluggish download time.

A number of technologies are available now that help with the bandwidth problem. *Broadband*—the term used to describe high-speed, high-capacity transmission channels—is being seen as the solution to slow and frustrating download times for space-greedy digital content, such as video-on-demand and multimedia communication. Among the more promising broadband technologies are the Digital Subscriber Line (DSL) and its relatives. The one we probably hear about most often is ADSL, or Asymmetrical Digital Subscriber Line. Ordinary telephone lines (often referred to as "twisted-pair copper wire") are used by ADSL but information is delivered at vastly faster speeds than ordinary dial-up modem services. This technology is important because it uses conduit that most people already have in their homes—the telephone line; no new wires and cables need to be installed. According to some estimates, DSL can deliver information to a home computer from the Internet up to 125 times faster than with modems. These high-speed connections are essential for people who use the Internet for advanced multimedia applications and exchange large files (e.g., for video, audio, graphic, real-time communication, etc.).

Many companies are now in the broadband business and trying to get your business. These companies include long-distance phone companies, cable operators, local phone companies, Internet service providers, and others. Various DSL services are marketed to residential customers, small businesses, and large corporations who consider broadband capabilities to be necessary or desirable. Qwest, for example, which not long ago merged with U.S. West, offers several DSL line speeds to its customers and promises dramatically quicker data transmission rates than without the service. (Keep in mind that typical modem speeds range from about 28.8 Kbps to 56 Kbps.) Examples of the increase in speed using DSL technologies[5] are as follows:

- Asymmetric Digital Subscriber Line (ADSL)—delivers up to 1 Mbps upstream (*from* the user) and more than 7 Mbps downstream (*to* the user).
- ISDN Digital Subscriber Line (IDSL)—provides a dedicated connection of up to 144 Kbps upstream and downstream.
- Symmetric Digital Line DSL (SDSL)—transmits up to 1.5 Mbps in both directions over a single twisted copper pair.

A number of existing line qualifications need to be met before customers can get DSL enhancement, and, perhaps more important, access to DSL service largely depends on where you happen to live. The big cities and primary metropolitan areas are better served. If these qualifications are met, DSL services can usually be ordered through a telecommunications provider such as Qwest or Verizon. Verizon allows customers to order the service online and even install it themselves using a self-installation kit. Costs for the service vary, depending on the speed of transmission desired and other factors. It typically ranges from about $40 to $200 a month. Flip through most technology publications, however, and a dozen or more companies may be advertising their broadband services.

Asynchronous Transfer Mode (ATM) is another technology that has been employed for the transmission of large quantities of information. It sends data as fixed-length packets or cells and can be used from desktop computers to local area networks and wide area networks. Asynchronous Transfer Mode allows existing telecommunications infrastructures to carry more data over their networks and is seen as holding promise for bandwidth-intensive services such as videoconferencing, video on demand, medical imaging, distance education, interactive multimedia games, and so forth, because of its ability to transmit voice, video, and data over the same network, as opposed to through separate networks (e.g., telephone, cable, and computer network).

The deployment of *optical fiber* (also known as *fiber optic cable*) as a transmission medium has had significant impact on the ability to move vast amounts of data around at speeds of light. Optical fiber is made from very thin and pure strands of glass through which information passes as pulses of light. Eventually, experts predict, most telecommunications conduit—the hard wiring—will be fiber optic. This is because of its far superior bandwidth capacity when compared to any other existing conduit, such as twisted-pair copper wire or coaxial cable, and the ability to transmit data digitally rather than in analog form. Anecdotally, one instructional video tried to convey the bandwidth capacity of optical fiber with this example: It said that the entire *Encyclopedia Britannica*, in digital form, could be sent from New York City to Philadelphia in under one minute.[6]

As of August 2000, Qwest Communications International had deployed more than 25,000 miles of optical fiber in North America. Other companies building an optical fiber network in the United States include Level 3 Communications and Aerie Networks.[7] As the Internet and digital media become more pervasive, optical fiber is seen as the conduit that can and will deliver the digital signals. Telephone companies already use optical fiber extensively, but signals travel on both optical fiber and copper wire—the latter especially as signals reach closer to residential homes, which are most likely wired with traditional twisted copper pair.

Satellite technology is another critical component in the emerging new media environment. Although originally used for government, corporate, and media communications, including the transmission of news and entertainment from one location to another for rebroadcast by cable and terrestrial television stations, satellites today serve the end-user more directly. With the popularity and growing adoption of direct broadcast satellites (DBS) and the greater use of satellites for personal hand-held communication devices, satellite technology is more accessible to the average consumer than ever before.

For a monthly fee and with the requisite equipment, companies such as DirecTV and EchoStar give subscribers access to hundreds of channels of entertainment, news, and information. Before the availability of DBS satellite dishes, achieving good-quality satellite reception was relatively cumbersome and complex for the average consumer. Satellite dishes were large (many feet in diameter) and not easy to install. They would have to be set up on rooftops or in back yards with cabling running from the large, heavy dish to the equipment inside the home.

Satellite Dish
Satellite dishes are a common sight at television stations throughout the counrtry. This dish, at a local public television station in Hawaii, reduces the problems that geographical distance can cause when sending or receiving information. Before the widespread use of satellites, news footage had to be flown to their destinations, creating delays and increased logistics.

Direct broadcast satellite dishes are about 18 to 24 inches in diameter and can be mounted to the side of a house. Because of their size and appearance, they are sometimes referred to as "pizza dishes." The monthly fee for DBS service varies, depending on the plan one purchases (in other words, how many channels one wants to receive). These days, there are so many channels available—in broad categories such as general interest, family/children, religious, educational, broadcast/superstations, movies, news and information, sports, music—that one could probably channel surf for hours without repeating channels too often. There seems to be something to satisfy just about anybody's interest.

Satellite broadcasting is considered to be the major competition to cable broadcasting. They share many of the same channels. But satellite reception is unidirectional—the programs are downloaded from a satellite in orbit. Future satellite systems will have broader functionality. Users will be able to upload information for wireless transmission. Experts disagree about how satellites will fare against cable TV and other broadband technologies, but many companies have already invested large amounts of money to create a dynamic satellite network that will feature a plethora of download and upload services. With the full convergence of hand-held computers and telephones, satellites may provide the means through which personal global communications is achieved.

Convergence

It has been predicted, and one can only speculate at this juncture, that most communication and information technologies in the home will converge into a single, ubiquitous system. (Whether the same company will control these technologies is a separate issue.) Hence, the telephone, Web access, e-mail, television, radio, home shopping, online banking, and more—all of these things will be digital based and coordinated through a central nervous system in the smart home. At one time this sounded like science fiction, but we may be moving closer to that reality as digital television (DTV) or computer/television hybrids (WebTV, PCTV) become adopted by a critical mass of U.S. consumers and homes become equipped to handle massive amounts of voice, video, and data transmission in digital form. This transmission may itself require a convergence of technologies—twisted-pair copper wire, optical fiber, coaxial cable, satellite broadcasting—as well as bandwidth-enhancing services (ATM, ADSL, ISDN, etc.).

Will the smart home, wired neighborhood, national information infrastructure, and global communication network be our run-of-the-mill digital media environment of the future? Or should we heed the warning of one conservative student of forecasting methods, Steven Schnaars, who argues that enthusiastic predictions of revolutionary change (brought on by technology) tend not to be realized if we look at examples throughout history. He asserts that "we should temper our forecasts, especially technological forecasts, to postulate smaller changes than we have in the past" (1989, p. 184).

Others argue just the opposite. George Gilder is known for his creative, some might even say hyperbolic, visions of the future. In a *Wired* magazine feature article, Gilder is portrayed as an unapologetic technological enthusiast, casting aside the cautious predictions of his colleagues and furthering a characteristically fanciful and optimistic view for what he calls the *Telecosm.* "Unknown entrepreneurs will invent new technologies to solve the current problems that hex Internet commerce—including encryptions, viruses, and nanobuck transactions," he is quoted as saying at technology roundtable featuring himself and other luminaries such as Alvin Toffler (*Future Shock*) and MIT economics professor Lester Thurow. Gilder continued, "The Internet will multiply by a factor of millions the power of one person at a computer" (Bronson 1996, p. 126). Gilder's Telecosm is an infinitely massive communication space enabled by an "an infinitude of potential bandwidth" and the "endless multiplication of spectrum use and reuse."[8] People such as Gilder see digital technologies significantly impacting everything from education to politics to personal relationships.

Will the public support Gilder's vision of a Telecosm? In other words, will they buy and use the technologies that these techno-visionaries believe will permeate our everyday lives? No one knows for sure. The S-Curve Pattern of Consumer Adoption (see Figure 5.1) graphically depicts the successful consumer adoption of a new technology.[9] Most communication and information technologies have gone through this pattern, characterized by a period of early adoption; followed by growing, gradual interest in the technology; followed by a sudden, rapid adoption of the technology; and finally entering into a stage called "saturation," where the vast majority of households have the technology. The telephone, radio, and television have all gone through these stages of consumer adoption, but the difficult-to-answer question is: When will the period of

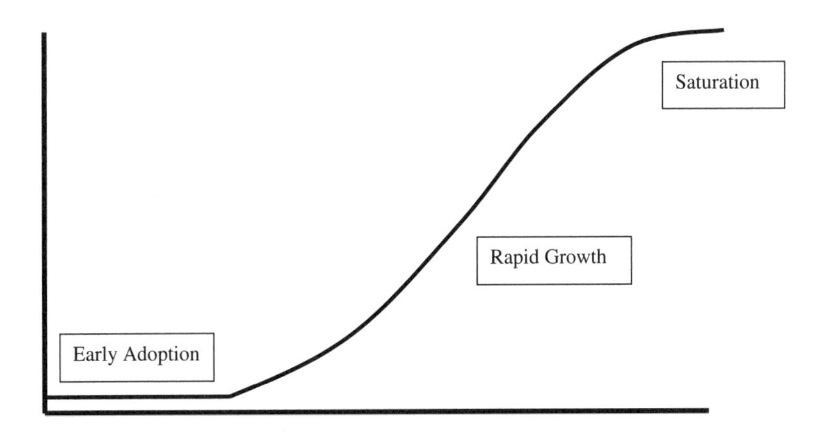

FIGURE 5.1 ***S-Curve Pattern of Consumer Adoption.*** The S-Curve Pattern of Consumer Adoption illustrates the stages that a new technology typically goes through if it is successfully adopted by the mass consumer market. When it is first introduced to the mass consumer market, a small number of people (early adopters) purchase it, even if it is unfamiliar and expensive. As the technology becomes more familiar and popular, it goes through more rapid adoption by the mass consumer market, until it reaches a stage where almost everyone has it in their household (saturation). Television, for example, went through these stages beginning from the early 1950s.

sudden, rapid adoption occur (if ever)? And, what are the factors that contribute to technological innovations being adopted by a critical mass of the population? These are questions worth pondering as we try to envision how society and our everyday lives will be affected by emerging technologies in our libraries, schools, businesses, and homes.

Unsuccessful technologies—such as the Picturephone, a video telephone introduced by AT&T in 1964 at the World's Fair and offered to residential consumers in 1970—did not catch on with the public for a variety of technological, economic, and social reasons. It was expensive, the existing telecommunications network was designed to carry voice not video, and the jerky quality of the picture left something to be desired. Besides, do people really want to see each other when they are talking on the telephone? Most people can think of at least a few reasons why they'd rather remain unseen. Other iterations of the video-telephone technology have been introduced to the mass consumer market since the 1970s and have, in a word, flopped. Today, with the emergence of broadband networks, wireless technology, the Internet, and hand-held computers, the video-telephone is making yet another comeback. It's anyone's guess whether it'll catch on this time around. Cost, ease of operation, quality of the technology, convenience, status appeal, value, and a host of other factors (including whether the technology is interoperable with other communication devices, such as a good old-fashioned "normal" telephone) all figure into the mass consumer adoption of a new technology.

One thing is for sure, however. Digital media cannot survive unless they are profitable. The global information infrastructure, while subject to government policies and regulations (see Chapter 7), is being developed by the private sector—from very big businesses (global multinational media and telecommunication companies) to very small businesses (two- or three-person companies in somebody's garage) and everything in between. The next chapter deals with the economics of digital media, including issues about corporate ownership and electronic commerce. Will digital media be the engines that drive the United States and global economy in the twenty first century? Many sectors of U.S. society are literally banking on it.

Questions for Discussion and Comprehension

1. Find someone who graduated from college before the mid-1980s and ask him or her about the level of technology integration in the curriculum. What were libraries like back then? How much were computers used? Were there computer labs on campus? What kind software applications were available, if any?
2. Do some research on the Blackburg Electronic Village (BEV). Why was this experiment such a revolution in online community formation? Is this a good prototype for other communities to use? What are the prerequisites for a community to be wired as Blacksburg, Virginia, is?
3. Discuss the future of technological innovation in your society. How do you think your society will change five years from now because of technology? How will this impact your own life? (*Hint:* Think about things such as commerce, health care, relationships, home shopping, etc.).

Endnotes

1. Press release by Nielsen//NetRatings, August 17, 2000, "Home Internet Access Reaches Critical Mass in the U.S." Nielsen//NetRatings is a partnership between AC Nielsen and NetRatings, Inc.
2. Press release by Media Metrix, August 9, 2000, "Women Outpace Men Online in Number and Growth Rate According to Media Metrix and Jupiter Communications." Unique visitors were defined as "the actual number of total users who visited the reported Web site or online property at least once in the given month. All Unique Visitors are unduplicated (only counted once) and are in thousands."
3. J. Martin McOmber, "Home Sweet Home for the Gateses," *Seattle Times*, September 11, 1997.
4. Robert Hawkins, "Don't Expect Every New House to Be a Smart House, Just Yet," *San Diego Union-Tribune*, October 20, 1998, p. 3.
5. These examples are from Qwest's own literature online. See http://www.qwest.com/smallbiz/dslTechnical.html.
6. Foundation for Advancements in Science and Education, *Digital Communication* (videotape), Los Angeles: FASE, 1995.
7. Andrew Backover, "Battle of the Bands: 3 Companies in Metro Area Network into the Future," *Denver Post*, August 7, 2000, p. F-1.
8. This is from an April 3, 2000 article Gilder wrote for a publication called *Forbes Global* and posted on his Web site.
9. The S-Curve Pattern of Consumer Adoption has been a focal point of presentations by John Carey, director of Greystone Communications, a longtime researcher of media technologies. He is the co-author (with Mitchell Moss) of an article, "The Diffusion of New Telecommunication Technologies," which appeared in the June 1985 issue of *Telecommunications Policy*.

6

Digital Media and the Economy

The Digital Economy

When Microsoft Chairman Bill Gates spoke at the Colorado Technology Summit hosted by Colorado's Governor Bill Owens in Denver, he said that "the digital economy is an unstoppable thing."[1] Attributing this unstoppable economy to the continued development digital media and broadband networks, Gates said, "The home will be different. Work will be different. Education will be different."[2]

Although Gates may have had his detractors some years ago, it is clear that digital media continue to impact the economy in important and sometimes even tumultuous ways. One of the most talked-about topics in recent years has been the emergence of a digital economy, a phenomenon that is as complex as it is vague. What does it mean to have a digital economy? What does it consist of?

This chapter looks at different components of a digital economy—namely, institutions, commerce, and technology. (The following chapter looks more closely at regulation.) The digital economy involves a complex web of business, trade, and technological relationships that cannot be easily summarized but is characterized by a robust information technology (IT) sector and a steadily increasing consumer market for digital media goods and services. The reason many analysts see promise in the prospect of a digital economy is reflected in the growth of the U.S. economy in the 1990s, especially during the second half of that decade.

According to U.S. Department of Commerce reports, the U.S. economy has enjoyed recurring bouts of good news in the past decade: Economic expansion occurred and showed no signs of slowing down at the dawn of a new millennium; labor productivity was high, doubling in the late 1990s; the country experienced its lowest unemployment rate in a generation; core inflation remained low; and investment in the IT sector boomed. The Department of Commerce attributed this period of prosperity in large part to what it called "the new economy."

In its third annual report analyzing the digital economy, the Department of Commerce found that rapid adoption of Internet technologies by U.S. consumers, explosive increase in innovation, dramatic cost reductions in computers and computer equipment, and growing interest among various sectors of society in electronic networking and commerce were crucial factors driving the growth of this new economy.

The Executive Summary of the report, *Digital Economy 2000*, ended on this optimistic note:

> [A] growing body of evidence suggests that the U.S. economy has crossed into a new period of higher, sustainable economic growth and higher, sustainable productivity gains. These conditions are driven in part by a powerful combination of rapid technological innovation, sharply falling IT prices, and booming investment in IT goods and services across virtually all American industries. Analysis of the computer and communications industries in particular suggests that the pace of technological innovation and rapidly falling prices should continue well into the future. Moreover, businesses outside the IT sector almost daily announce IT-based organizational and operating changes that reflect their solid confidence in the benefit of further substantial investments in IT goods and services. The largest and clearest recent examples come from the automobile, aircraft, energy and retail industries, which all have announced new Internet-based forms of market integration that should generate large continuing investments in IT infrastructure. These examples mark only the beginning of the digital economy. (U.S. Department of Commerce, 2000)

These hopeful signs signaled the dawning of a new era for the U.S. economy that would entail dramatic changes in the way that firms do business with their publics. Digital media change business relationships in ways that can both benefit and hurt firms' productivity. For example, a small business that sells ceramic pots from its street-front store in Eureka, California, may well benefit from using the Internet to market their pots to a global audience, accepting orders and payments online, and delivering the pots via the U.S. Postal Service or an overnight courier.

On the other hand, a travel agency in Evanston, Illinois, may find its business dwindling because travelers are finding it more convenient and quicker to purchase their tickets online directly from the airlines or through an online discount broker, such as Cheap Tickets (http://www.cheaptickets.com), Airlines of the Web (http://www.flyaow.com), MSN Expedia (http://www.expedia.com), Lowestfare.com (http://www.lowestfare.com), and a host of others. According to a report by the investment bank Bear Stearns, the travel industry is expected to experience massive growth in the next few years, but the number of traditional travel agents will decrease. The report said that about one-quarter of travel agents will lose their jobs as airlines cut travel agent commissions and sell tickets directly to customers over the Internet. In the first six months of 1999, the report said, 1,800 travel agencies had already gone out of business.[3]

It is too soon to say whether this trend will continue in the long run or apply to many other kinds of businesses.

A More "Equal Playing Field"?

Some argue that a digital economy "levels the playing field," allowing smaller firms to compete with larger ones without having to invest large amounts of money into capital, such as retail space, furnishings, and office equipment. They say that the digital economy is what permitted a company such as Amazon.com, for example, to enter a well-established market and successfully compete with traditional book distributors. Other examples of "online companies" that compete or at one time competed with their traditional counterparts off-line include online pharmaceutical companies (e.g., http://www.drugstore.com, http://www.planetRx.com, or http://www.rx.com), grocery stores (e.g., http://www.webvan.com, which bought out its competitor homegrocer.com in September 2000), pet stores (http://www.pets.com), video and music stores (e.g., http://www.kozmo.com), florists (e.g., http://www.1800flowers.com), toy stores (e.g., http://www.etoys.com or http://www.KBkids.com), and a long list of other online companies (dot-coms) that are trying their hand at electronic commerce. Some of these companies have since been devalued or have gone bankrupt. Nevertheless, in the period from April to June 2000, consumers spent $5.5 billion shopping on the Internet, an increase of 5.3 percent from first quarter (January to March 2000) sales.[4]

It is true that in a digital economy, the barriers to entry are easier to cross over than in the old economy. That is why college students or recent college graduates—newcomers, virtually "nobodys" in the corporate world—with great ideas and high-risk initiative can build up an online company, attract considerable venture capital, launch their business amidst excitement and fanfare, and become instant millionaires if their "make or break" enterprises catch on with the public.

The new economy is what allows a Steve Case, founder and chairman of America Online (AOL), which only existed since 1985, to buy out the largest media conglomerate in the world (Time Warner) for about $180 billion. Case was 41 years old when the deal was announced in January 2000 and only in his mid-20s when he started AOL. Jerry Yang, the founder of Yahoo! was age 27 when he took that company public in 1996. Yahoo! was valued at approximately $70–100 billion in late March 2000 but, like other dot-coms, has since drastically fallen in value. Of course, Bill Gates's personal wealth is now the stuff of legend. It has gone up to the $90 billion mark in the past, depending on the whims of the stock market and his philanthropic generosity at the moment. These are extreme cases, obviously, but they reflect the popular perception of how successful dot-com entrepreneurs can become at the right place and time. And even though dot-coms have experienced a sudden downturn in their value and profitability, it is nevertheless true that hordes of young people have already become incredibly wealthy in a relatively short period of time during the more prosperous years.

Inspired by these breathtaking stories of quick wealth and corporate power, many people today are rushing toward the digital economy, hoping to partake in the largesse that may await them as well. But before the picture gets too rosy, the success stories need to be put into context.

First, although the digital economy and e-commerce is definitely growing in the United States, the emerging online business sector is highly unstable. The same report that predicted the number of traditional travel agents would decline also predicted that only about one-fifth of the online travel companies would survive years into the future. The same might be said for other online businesses. Many have jumped into the competition and hope to exploit the growing and undertapped online consumer market, but not all will survive. In fact, for all the successes focused on by the news media, many more are struggling to stay afloat and others have long since died. Clearly, there are both winners and losers in the dot-com world, and those businesses that quietly go belly-up should serve as a reminder that the digital economy yields its share of wounded and fatalities, just like in the off-line world.

Second, without discounting its growth and promise, online sales must be kept in proportion. Certainly, $5.5 billion sounds like a stunning amount of money to those without the big picture, but the reality is that total retail sales during the second quarter of 2000 totaled $815.7 billion, which means that the online proportion comprised not even close to 1 percent of the total.

Third, technology stocks have a mixed reputation in the investment world. As any stock market investor can attest to, technology companies, particularly start-ups, are risky and volatile. The second week of April 2000 saw a 25.3 percent plunge in the technology-dominated Nasdaq composite index and set off wild concerns about what would happen next now that the bubble had burst. Technology companies saw their stock prices plummet before their eyes in a matter of days and sometimes hours. One newspaper described Wall Street this way on April 14: "The tickers on trading floors across the city were bathed in a sea of red—the color that announces declining prices. The blood bath was most dramatic in the technology area as the momentum selling in overvalued high-tech stocks continued."[5]

This catastrophic week on Wall Street was optimistically called a "correction" by many Wall Street analysts after the fact. The truth is, no one was sure where the market would head next. In news reports immediately after the event, many references were made to the stock market crashes of October 1929 and October 1987. As it turned out, much to the relief of Wall Street, the market recovered.

The *New York Times* reported the next week,

> "Friday's largest-ever point loss on the Nasdaq stock market was followed yesterday by the largest-ever point gain—putting to rest, for now, fears of a meltdown on Wall Street. No one could know before trading started yesterday morning whether stocks would continue to plunge or, as has become the rule in recent years, recover as investors swooped in to buy what they perceived to be newly undervalued stocks. By the same token, no one can know if yesterday's snapback will last. By historical standards, stock prices still remain high."[6]

The recovery continued into the rest of the year, but the lesson was not to be forgotten. In the back of everyone's mind was that ominous week in April when the bottom fell out from under Wall Street.

Sure enough, in April 2001, the Nasdaq was again in a slump, fueled by one tech company after another reporting disappointing earnings. On April 4, 2001, the Nasdaq was down an alarming 32.3 percent for the year. A *New York Times* headline referred to the combination drop in Nasdaq and a possible decrease in consumer spending as the "Nightmare on Wall Street." The news was dismal:

> Technology shares continue to tumble as demand wanes for Internet routing equipment, cellular telephones and fiber optic cable. Fears of rising bankruptcy filings among smaller telecommunications concerns have infected industry giants like Nortel Networks, Lucent Technologies and Qualcomm. All three hit new 52-week lows yesterday.

> In the last two days alone, the Nasdaq composite index is down more than 9 percent, closing yesterday at 1,673. It has lost more than two-thirds of its value since it reached a close-of-trading-session peak of 5,048.62 on March 10, 2000."[7]

The dramatic ups and downs in the stock market have made many investors wary of technology stocks altogether. In the midst of this economic bloodbath, many financial analysts have explained that technology stocks were overvalued to begin with. A company's *value* does not mean that the company is highly profitable, or is making any profit at all for that matter. When the upper management of a start-up technology company decides that it wants to issue stock for the first time, the process is called *taking the company public.* It does this by allowing investors to purchase shares in the company (i.e., to buy stock in the company) through an *initial public offering* or *IPO.* It may decide to do this because it (1) believes that its product or service will be met by an enthusiastic market and (2) wants to "grow" the company and needs funding beyond what venture capitalists (investors who are willing to risk investing in a new firm but who also have a say in how the company is run) are able to provide.

The company needs to find investment bankers who are willing to buy the public shares at a set price and then resell those shares to the public. In other words, they underwrite the cost of the shares. The incentive for the investment bankers is to make a profit in the reselling. Members of the public who are interested in purchasing shares in the company have the right to read a company prospectus, a legal document prepared by the company and its underwriters. The prospectus has detailed information about the company's management, history, financial health, products and/or services, and so forth, including an assessment of any risks the company faces. Before the stocks go public, the underwriters decide what they will pay for each share.

There are many reasons why stock prices rise and fall. If anyone could accurately predict why and when these rises and falls would occur, he or she would be in great demand by investors. One reason stocks fall is that investors begin to believe that the shares they hold in a particular company are overvalued or will soon lose their value, so they sell them off while they can to avoid losses.

The great irony in a digital economy is that companies can be valued at billions of dollars and not even have turned a profit yet. A company may have a stockpile of money in reserve, thanks to enthusiastic investors, and that is reflected in the company's market value, but if it burns those cash reserves and is not replenishing it through revenue generation or the attraction of other investors, the company is in serious danger of eventually going under for its inability to meet its liabilities. Of course, a company, especially a new company, does not have to be profitable to have high market value. It could be the "first mover" in a new market that has tremendous potential in the eyes of investors. It could be that investors are perfectly willing to wait for other conditions to emerge (e.g., greater adoption of Internet technology) before expecting to see company profits. Some of the more cynical observers of technology stocks, however, consider many of them to be contributing to a bubble economy just waiting to burst; perhaps April 2000 was merely a precursor?

Part of the risk of the stock market (and for some, perhaps, the allure) is that it is often so unpredictable. Announcements and predictions of various kinds from government, analysts, or the companies themselves can send stock prices reeling in either direction. The market is sensitive to rumors, tentative projections, speculation, and predilections. Major investors cling to every word uttered by Federal Reserve Chairman Alan Greenspan, trying to read between the lines and foretell how his words will affect stock performance and investor confidence down the road. These days, investor confidence means more than just the individual investor (i.e., a single person). In this age of colossal institutional investors (mutual fund companies, insurance systems, banks, universities, retirement programs, etc.), the volume of stock trading can be massive on any given day for a particular stock. Panic selling can potentially affect the entire stock market. With such high-volume traders closely monitoring the stock market and reacting to subtle clues and tip-offs about the economy and stock performance, it's no wonder stock prices can plunge and recover at such erratic and high speeds.

Of course, all of these things preceded the terrorist attacks on the World Trade Center and the Pentagon on September 11, 2001, and the ensuing fears about more terrorist attacks and actual cases of anthrax-related illnesses. In addition to the thousands of innocent lives lost, the stock market and U.S. economy severely suffered after this tragic event, the long-term effects of which have yet to be seen as of this writing.

Finally, for all the feature stories about Microsoft and companies producing 30-something-year-old millionaires like there was no tomorrow, many people in society will not benefit from what the new economy has to offer. In fact, for lack of skills, access, resources, or interest, these people may find the gap between themselves and the techno-literate class ever widening, leading to the much-touted "digital divide." The digital divide will hurt some people more than others. Wealthy and powerful people without technical skills can hire or persuade others to attend to their lacks, as they always have, but poorer and historically marginalized people may be left behind to fend for themselves unless educational programs and services successfully resolve the inequities.

Hence, the promises and perils of the digital economy need to be considered side by side. As government and business leaders invoke a long and prosperous future for the digital economy, it is still not clear who will stand to gain and lose the most as digital media and broadband networks become more routinely part of trade and commerce. Some believe the global conglomerates will wrestle control of the new economy away from the "little guys" as time goes on, perhaps through mergers and acquisitions or through competitive business practices that push the envelope of business ethics. Others think the lords of the old economy will grow increasingly out of touch with the rapid changes going on around them and be overtaken by the young upstarts. Who is right? The answer is probably that both will occur. The following section looks at some important players, developments, and issues in the digital economy.

Institutions, Commerce, and Technology

For all the ink spent on writing about the digital economy in the past few years, the reality is that the benefits and detriments of this economy have yet to fully play themselves out. As mentioned earlier in this chapter, a large number of institutions have much to gain and lose from a digital economy: retail outlets, banking, education, advertising, telecommunications, entertainment, media, and so forth. What are these institutions doing to prepare themselves for the future?

At a minimum, more and more businesses are carving out a presence for themselves on the Web by creating their own Web sites. Compared to advertising in the traditional media (television, radio, newspapers, and magazines), creating a Web site can be a relatively inexpensive endeavor. Of course, the fancier and more sophisticated sites may cost more, but basic Web sites are generally affordable even for the most modest of businesses. Someone within the business organization may be able to build it; if not, one can usually find an eager student or small company to do it at a low cost. With some initiative, however, most people can learn to build a basic Web site on their own, as well as find out how to have it hosted on a Web server for a monthly fee (if it is an outside Web server rather than one that exists within the company).

Having a Web site for one's business is hardly anything to get excited about these days. The more important question is: What do we do with it? That's where good old-fashioned creativity and a strong marketing plan come into play. Web sites can serve many purposes. They can be purely informational—for example, a restaurant might want regular and new customers to visit its Web site and check what the specials of the week are going to be, to find out hours of operation without calling, or even to learn something about the history of the establishment, its management, and those who do the cooking. More information might be offered about certain types of herbs, oils, spices, and other ingredients.

Web sites can go beyond the informational level, however, and become more interactive and conversational. That same restaurant might have a part of its site devoted to visitors' comments, where visitors can post messages to the chef or management about their experiences at the restaurant, what they liked and didn't like, what could be improved, and so forth. (Only the bravest of establishments would allow a discussion area like this to exist unmoderated!) Those who post messages could follow up on previous comments, agreeing or disagreeing with those who "spoke" earlier. There could be a "recipe swap" of some kind, or suggestions on what to order for those on food restrictions.

Beyond the interactive and conversational level, there could be the e-commerce level. Goods could be sold online—bottled sauces, cooking aprons, coffee mugs, to name a few. Visitors to the site could order these items for themselves or as gifts for others and pay for the merchandise online. If the site should become very popular, one might even sell advertising space to interested parties as an additional source of revenue.

Obviously, the more things one chooses to do with a Web site, the more skilled labor and sophisticated technology it will require. The question any business has to ask itself is: Is it worth it? When factoring in the cost of labor, technology, and special taxes or fees, the online business model may not be particularly feasible or wise. But even if a site is not directly revenue generating, it may still have value to the company. As a mechanism for drumming up new business or reinforcing customer satisfaction and loyalty, the site may be well worth the expense in time and capital.

Johnson and Johnson's Web site (http://www.jj.com) is a good example of a company Web site that serves a wide variety of purposes. As an informational site for potential investors, it provides detailed investor relations data, including its stock quote, financial reports, and trading history. It also contains information about getting a job within the company, news about the company and its products, and links to sites on baby care and women's health. Although this site may not rake in money through the sale of goods and services online, it serves a purpose that has less to do with e-commerce and more to do with traditional off-line commerce. Johnson and Johnson, as a leading manufacturer of personal health-care products, can build strong customer relations by providing them with information about breastfeeding, skin irritations, and diarrhea. Products to deal with these things can be purchased from a conventional retail drugstore. This kind of site is not engaging in e-commerce but is driving business from the Web site to traditional retail outlets.

Web sites that are more pointedly in the realm of e-commerce are those that try to sell you something online, collect your money online, and distribute the goods online or through the mail. Book sellers Amazon.com, Barnes and Noble, and Borders all have Web sites where you can locate books online through browsing or search engines, put them in your online "shopping cart," and purchase them with a credit card or with an online gift certificate that someone may have sent you via electronic mail for your birthday. As of this writing, Amazon has partnered with Borders to serve as its online distributor. The Barnes and Noble Web site (http://www.bn.com) sells regular hard-copy books, electronic books (you need to download an eBook reader software), college textbooks, out-of-print books, music, DVDs and videos, software, prints and posters, magazine subscriptions, and even online courses. This last category—online courses—may come as a surprise to some, but it is a rather ingenious idea. Visitors to the Web site can register for courses on a wide variety of topics from the highly technical to more leisurely pursuits such as travel and lifestyle. Although the courses themselves are free, the course materials (generally books) are not, but they can be conveniently purchased online right on the very Web site that you are taking your course on. Very convenient. Very clever marketing strategy!

The brains behind this endeavor used to be a company called notHarvard.com, which changed its name to Powered, Inc. It is in the business of what it calls "eduCommerce," which it defines as "The next big thing. Using free education as a powerful customer acquisition tool—enhancing your customer value proposition. Free education as a sales and marketing weapon to drive greater stickiness, deeper customer intimacy, and higher brand loyalty resulting in incremental revenue. Because sellers need to teach and buyers want to learn."[8] Some may find this philosophy rather crude and exploitive, but in the digital economy, the rules often get changed. Education is no longer the exclusive purview of traditional educational institutions but could be offered by educommerce corporations. That is why so many traditional universities that would have at one time shunned the idea of Web-mediated distance education are now embracing the opportunity, charging students for online credit hours and issuing certificates and even college degrees for a completed assortment of coursework. If they don't take advantage of the Web as an educommerce tool, someone else will. Distance education courses in many (not all!) circumstances can be more accessible, cost efficient, and as effective as their on-campus counterparts. If universities don't start jumping on the distance education bandwagon, they could be left behind carrying an empty money bag.

Web-based distance education is just one example of how traditional industries are struggling to find their place in the digital economy. There is a long list of other examples. Financial transactions online are at the heart of the digital economy. Money flowing electronically over long distances is hardly a contemporary phenomenon.

Banks have "wired" funds from one place to another for many decades, and the networked financial system of charge cards, debit cards, credit, and automatic teller machines (ATMs) have been in place for a long time. So-called electronic money existed long before the emergence of the World Wide Web. The difference, however, is that the Web is now becoming a significant medium for more personal—rather than institutional—financial transactions. Electronic trading in the stock market presents one useful case study.

At one time, long before online trading, if you bought 100 shares of stock in the hypothetical company Gumshoe Industries, you would need to call a broker at a brokerage house (i.e., an investment firm), such as Morgan Stanley Dean Witter or Merrill Lynch, and open an account. You would then place an order through the broker, who would act as your agent, to buy 100 shares of Gumshoe Industries at a specified price, say $15 a share (the asking price). The broker would execute the order on your behalf. The cost to you would be $1,500 for the price of the stock, *plus* you would have to pay your broker a commission, which would be split between your broker and his or her brokerage house. These commissions could be quite high, depending on the level of service provided. Full-service brokerage houses theoretically provide a high level of service, doing research for you on companies and making recommendations and providing financial advice. You would receive a bill in the mail for the total amount owed, and after you paid it you would receive some written confirmation of your stock ownership.

Unfortunately, not everyone benefited from or needed the same level of service as everyone else, but people had little choice except to pay the commissions regardless of what they actually received in terms of service. Small-time investors who didn't have a lot to spend on stocks may not have received as much attention as their more affluent counterparts, but they paid the same commission anyway. In the mid-1970s, less regulation of the investment industry led to the appearance of discount brokerages. Discount brokerages, such as Charles Schwab and Co., were a welcome relief to those investors who could do their own research and didn't need (or get) financial advice from their brokers and simply needed someone to execute a transaction. Discount brokers charged considerably lower commissions, which reflected the level of service they provided.

Now you can log on to the World Wide Web, go to any one of a number of online discount brokers, and open an account by filling out an online form. There is some paperwork that needs to be handled through the mail, but once that is completed, the rest of the work is done online. E*Trade is a popular online discount broker with a comprehensive Web site that offers a full range of information from educational articles for the beginning investor to complex stock reports. Investors can check their stock prices online, buy and sell stocks with a few clicks of the mouse and keyboard, and check their personal accounts right from the Web site. Commissions are negligible compared to full-service brokerage houses. The amount of information the online stock investor has at his or her fingertips is comparable to what only "professional" financial advisers used to have access to. Nonprofessionals, if so motivated, can develop their own skills and make (or lose) money through much more direct methods of buying and selling shares of stock electronically.[9]

An online discount brokerage firm like E*Trade is an excellent example of how the digital economy permits consumers to bypass traditional interfaces and conduct business in a more direct and efficient manner. Responding to competition, full-service brokerage houses are beginning to scale their fees to fit their services, but they are coming into the game late. (Full-service brokerages are not really in danger of going out of business, however, because of their ability to manage assets for very large investors.)

Nevertheless, brokerage houses are a good example of how the digital economy is changing the face of business in the United States. As e-commerce continues to grow, traditional business institutions will have to adapt to the new economy or risk becoming supplanted by digital services that are cheaper and more efficient for the consumer.

Increasingly, traditional mass media institutions have had to face the prospect of an "adapt or die" scenario. Most newspapers and television stations in the United States have sprouted online counterparts, and all of the media conglomerates have "new media" divisions that are trying to discover what kind of digital content will be most marketable to the consumer public and what business model will make the sale and distribution of this content most profitable. Media have a particularly promising future in the digital economy because their content can be digitized and sent through computer networks such as the Internet. This is not the case for, say, automobile dealerships or even online auction sites such as eBay, which can do a certain amount of business online but must still rely on traditional distribution channels to deliver the goods. Digital media content, however, can be created, sold, bought, paid for, and delivered through the very technologies that facilitate the digital economy. Computer software is another commodity that shares this practical quality.

An assortment of business models is currently under development throughout the traditional mass media industry. Existing mass media sites (both news and entertainment, with few exceptions) are notoriously unprofitable, but business executives are hoping this will eventually change since so much effort and money have already gone into developing their new media divisions. There seems to be a feeling that the formula for financial success is out there; they just have to discover what it is. Clearly, there are very high expectations. Online commercial pornography sites are often said to be the one sector of media entertainment that has actually been profitable using a paid subscription business model, but it is unlikely that most traditional media organizations will marshal their resources toward producing that kind of content. The reality may be that despite the exponential consumer adoption rate of the Internet, the number of users must be even higher still for large enough consumer markets to exist for certain kinds of goods and services. This is the point that we are at today.

The Importance of Networks

Perhaps nothing will propel the digital economy faster than the development of global computer networks, that extend to businesses, homes, schools, government, and civil society. The IT sector differs from most other industries in that its value and robustness increase as the number of people who adopt IT technologies and systems increases. This is known as *network effects.* A U.S. Department of Commerce report explains this phenomenon well:

> The more the technology is deployed, the greater its value. Compare certain information technologies to automobiles. When you own a car, its value to you is basically the same whether 5,000 or 1 million other people own the same brand of automobile. When you buy a computer operating system or graphics program, its value to you increases as more people buy it, because their purchases of the same program increase your ability to digitally communicate and interact. As these forms of innovation spread, the productivity benefits may increase at a faster rate than simply arithmetically. (U.S. Department of Commerce, Economics and Statistic Administration, 2000)

On a more proprietary level, the concept of *network effects* sheds light on why a merger like the one between AOL and Time Warner is so significant. It combines the massive programming content base and cable network of a traditional media conglomerate with the online network of the world's largest commercial Internet service provider. Theoretically, the larger the web of network connectivity in this mega-multimedia conglomerate, the greater value each individual consumer connection holds because of the corresponding access to massive varieties of content, services, and other people. In reality, the merger has been a financial drain on the company, which lost more than half its value in the year or so after the merger was announced.

The rise of a digital economy raises as many questions as it provides potential answers for a high-growth economy. Laws and regulations governing digital commerce are still in formation. It is not clear how to treat digital goods and services when considering tariffs and domestic taxes, for example. States are considering local sales taxes as most already have in place for tangible goods. Wildly popular online auction sites such as eBay, self-described as "the world's online marketplace" that links tens of thousands of people engaged in online transactions of one kind or another, raise new legal and regulatory questions. A Federal Trade Commission (FTC) attorney was quoted in the *Chicago Sun-Times* as saying that "the number of complaints about online auctions increased from 107 in 1997 to 10,700 last year [1999], out of roughly 19,000 complaints of general Internet fraud in 1999."[10] In addition to the FTC, the FBI has also investigated a range of illegal or questionable online activities.

As this chapter has tried to demonstrate, the digital economy is not any one thing but a web of related components—institutions, technology, and commerce—that together contribute to an exchange of digital goods and services or help support this exchange in some way (e.g., as advertising of digital goods and services in traditional mass media or the sale and distribution of networking hardware to facilitate digital media networks). Few would argue that the digital economy will continue to grow and influence people's personal and professional lives in a variety of ways. The next chapter looks at how the technologies around which such an economy rises should be governed and regulated.

Questions for Discussion and Comprehension

1. What does it mean to have a digital economy? There were times in history when the economy was based largely on agriculture, heavy industry, and service jobs. How is the economy changing (or not changing, for that matter) in the twenty-first century?

2. What kinds of companies could be threatened by the Internet if they don't have an online service component? How can the Internet "steal" business away from these companies, bypassing the "brick and mortar" storefront, so to speak, to provide goods and services directly to the consumer?

3. Do some research on the Web about the future of dot-coms and Internet-related companies. What do you think the future holds for this industry? What does it take to succeed? What is the current state of Nasdaq? Analyze the current and future health of the digital economy.

Endnotes

1. Jennifer Beauprez, "Gates Touts Digital World in Five Years: Software Titan Challenges Colorado at Tech Summit," *Denver Post*, June 21, 2000, p. C-1.
2. Ibid.
3. Salina Kahn and Donna Rosato, "Web to Cut Travel Agents by 25%," *USA Today*, April 18, 2000, p. 1-A.
4. Based on Nielsen//NetRatings and U.S. Commerce Dept. figures. Reported in ECommerce Times.com, Sept. 1, 2000. http://www.ecommercetimes.com/news/articles2000/000901-1.shtml
5. Sharon Walsh and John M. Berry, "Dow, NASDAQ Enter Free Fall," *Washington Post*, April 15, 2000, p. A-1.
6. "Rebound, and Relief, in the Markets," *New York Times*, April 18, 2000, p. A-24.
7. Ibid., April 4, 2001, p. C-1.
8. From http://www.notharvard.com.
9. To be clear, online investors are still going through a brokerage house and not buying directly off the stock exchange through computer networks. But the process for purchasing stocks is expedited by bypassing the traditional broker and entering orders directly from your computer.
10. Christine Hanley, "FBI Probes Bidding Fraud on eBay," *Chicago Sun-Times*, June 8, 2000, p. 24.

7

Government Policies and Regulations

Rules and Regulations

The control of communication channels and content by a society's social and political elite is a well-worn practice that probably dates back to antiquity. Ancient libraries, for example, were usually housed near or even within royal palaces and temples. Here, the storerooms of accumulated knowledge could be vigilantly guarded and restricted, if necessary, and information wouldn't get into the wrong hands or—worse—*minds.* The Catholic Church during the Medieval period, while responsible for the preservation of many manuscripts that might otherwise have been lost or destroyed, also jealously protected its intellectual largesse. By keeping a tight reign on manuscript reproduction in its monasteries and scriptoria, where monks meticulously hand-copied books and other documents as a form of spiritual labor, the Church could control what got into the hands of the public and what remained locked behind the walls of religious and political privilege. With a scarcity of religious texts available, the masses (who were mostly illiterate anyway) had to rely on the Church for its interpretation of the Scriptures and canon law.

At its political epoch, the Catholic Church wielded broad and intoxicating authority as judge, jury, and executioner—until Johannes Gutenberg's printing press and the Protestant Reformation undermined the Church's narrow monopoly on knowledge and social power. Even after the Reformation, the Catholic Church remained a forceful arbiter of what words could and could not be consumed by its obedient followers. Under threat of excommunication or other punishment, parishioners were forbidden from reading an assortment of books and documents that were antithetical to Church teaching. Moreover, scientific pioneers, such as the Polish astronomer Copernicus and Italian scientist Galileo, were soundly condemned for their theories and observations that presented a much different view of the cosmos than what the Church then believed to be the case. When Henry VIII successfully challenged papal supremacy in sixteenth-century England, however, he merely substituted his own brand of censorship for the one he conquered. Folkerts and Teeter (1989) write:

> To maintain his control, Henry imposed prior restraint, a system of prepublication censorship that has had many imitators over the years. When publications eluded the network of censors, Henry's government punished those responsible for "seditious libel." Such punishment could be both deadly and unutterably cruel. In 1529, Henry published a list of prohibited books, and on Christmas Day of 1534, he ordered printers to secure royal permission to operate. (p. 8)

Restrictions on speech and publication were the norm in the Western world until they began to be fought with reason and rebellion. John Milton's famous essay, "Areopagitica," delivered to Parliament in 1644, was a persuasive argument against requiring English printers to apply for printing licenses from the government. This essay, as literary as it was political, underscored the importance of free speech and helped set the philosophical foundation for reasonable—not arbitrary—communication law. Let "Truth" and "Falsehood" grapple, Milton pleaded, in a field of free expression. Truth, he believed, would percolate to the top and become evident without the meddling hand of government. "For who knows not that Truth is strong, next to the Almighty?" Milton wrote. "She needs no policies, nor stratagems, nor licensings to make her victorious; those are the shifts and the defences that error uses against her power."[1]

Clearly, the practice of trying to regulate communication content and channels have deep roots, as do attempts to stave off regulation in favor of free, unfettered expression. These countervailing forces are balanced differently in different countries, but in the United States the tradition since postcolonial times has been to vigorously protect free speech and a free press. The strength of this conviction comes out of bitter struggles the American colonists endured while living under the thumb of a controlling British monarch. The burgeoning free press in the Colonies contributed much to foundations of democracy as it renounced autocracy and encouraged thoughtful public discourse and deliberation, including that which may have been displeasing to the governing authority. When the First Amendment to the United States Constitution was adopted in 1791 as part of the Bill of Rights, it validated these hard-won battles to overthrow the repression of arbitrary power. So much of contemporary communications law is based on or is in some way related to these monumentally important words:

> Congress shall make no law respecting an establishment of religion, or prohibiting the free exercise thereof; or abridging the freedom of speech, or of the press; or the right of the people peaceably to assemble, and to petition the Government for a redress of grievances. —The First Amendment to the U.S. Constitution

Five fundamental rights are embedded in this concise statement—speech, religion, press, peaceful assembly, and petition—but it is the so-called free speech and free press clauses that are most relevant to communications law. For the most part, the print media in the United States have enjoyed the most First Amendment protection in comparison to the broadcast media—although certainly books, newspapers, and magazines have suffered their share of meddling government regulators. Radio and television, because of their use of the electromagnetic spectrum to transmit aerial signals over a limited number of radio frequencies, are considered to be users of a finite public resource. It is the federal government, through an administrative agency known as the Federal Communications Commission (FCC), that allocates this public resource. The FCC has created rules and regulations pertaining to the electronic broadcast media since the 1930s.

History of Federal Regulations

In the century preceding the growth of commercial radio, a number of important telecommunication technologies were invented, including the telegraph, the telephone, the wireless telegraph, and amateur radio. Inventors and hobbyists invested much time, money, and energy into developing radio technology in the early 1900s, and it soon became apparent that interoperator interference would ensue without an orderly allocation of radio frequencies. The U.S. Congress passed the Radio Act of 1912, which required radio operators to apply for a license from the U.S. Secretary of Commerce and Labor. Essentially, anyone who applied for a license got one. Because of bureaucratic inefficiencies and logistical imprecision, the licensing process was ultimately an ineffective attempt to manage the distribution of radio frequencies. As international tensions mounted in the period leading to World War I, the U.S. military grew increasingly anxious about the disorganized state of radio communications.

At the behest of the U.S. Navy, the government banned amateur radio operations during World War I, restricting the airwaves for military and government communications. After the war ended, however, nonmilitary radio operations started up in even greater numbers than before. From that point onward, the radio waves were in varying degrees of chaos and disarray. Congress finally passed the Radio Act of 1927, which established once and for all that Congress could regulate the use of radio waves in the public interest. The Federal Radio Commission (FRC) was created to regulate radio communication. Later, the Federal Communications Act of 1934 established the Federal Communications Commission, a powerful federal administrative agency still in existence today. The FCC and the 1934 act have been the basis of many rules and regulations pertaining to radio (and later television) in the United States for almost 70 years.

Digital Media and Regulation

In the more than six decades between the Federal Communications Act of 1934 and the Telecommunications Act of 1996, which effectively overhauled the 1934 law, the communications environment in the United States had changed dramatically. As is often the case, both at the national and international levels, laws and regulations lag behind technological development. Although laws can be created or modified to anticipate or preempt problems, they often arise in *response* to emerging problems where communications technologies are concerned. The regulation of radio frequencies by the FCC is an example of this, but there are many other examples that can be cited. Concerns about copyright, libel and defamation, invasion of privacy, the ability to access government information, the availability of obscene and indecent content, false advertising, and so forth, with evolving communication technologies have led to a substantial body of communication law that have filled entire textbooks and constituted a full course offering in itself. Publications on mass media and telecommunications law and, more recently, "cyberlaw" or "netlaw," are currently available for an in-depth treatment of traditional and emerging communication law. This chapter will provide an overview of key points related to the regulation of digital media.

By the time the 1990s rolled in, it was clear to lawmakers, communications industry representatives, and others that the omnibus Federal Communications Act of 1934 was growing increasingly outdated and would soon be obsolete and potentially obstructionist to a robust telecommunications industry in the United States. It would become obsolete because technological advances, particularly due to digital media, were creating an entirely new communications environment—wholly unlike the communications environment in the 1930s, when radio was the "new media" and existing communications technologies included the telegraph and telephone. In those days, *cable* didn't refer to television but to undersea telegraph lines. By the 1990s, the playing field had been turned upside down. For example, telephone companies were interested in offering video services. Cable companies were interested in offering telephone services. The regulatory wall erected in 1984 between providers of long-distance telephone service and regional telephone phone service was being eyed by many corporate players for eventual, albeit considered, disintegration. In general, the telecommunications industry was aiming for a more deregulated communications environment.

In addition, the Internet was fast becoming a popular medium, adopted by ordinary people and not just researchers and scientists. Without changes, prohibitive old laws would likely obstruct the ability of telecommunication firms to expand into other areas of service. Telecommunications firms, in particular, which were increasingly becoming tied into a wide range of Internet services, were concerned about the effects of long-standing regulatory hurdles.

It is certainly not the case that communications law remained stagnant from 1934 to 1996. Many laws went into effect during this time to accommodate the changing communications environment, such as the rise of cable television. A series of cable television acts beginning in 1984, for example, dealt with issues such as noncommercial community access programming, community jurisdiction to negotiate cable franchise contracts with potential cable operators, privacy of cable subscribers' records, cable rate increases, the relationship between broadcast operators and cable operators, and so forth.

Another major telecommunications regulatory development occurred when U.S. federal court Judge Harold Greene ordered the break-up of telephone giant AT&T in response to concerns about market monopolization. This order, known as a *consent decree*, went into effect in 1984 and gave AT&T the right to continue offering long-distance telephone service, but local phone service had to be provided by newly formed regional Bell operating companies, or RBOCs, which were prohibited from entering into the long-distance market. The original RBOCs were Ameritech, Bell Atlantic, BellSouth, NYNEX, Pacific Bell, Southwestern Bell, and US WEST. The consent degree was terminated by Judge Greene not long after passage of the Telecommunications Act of 1996.

Although the Internet had existed in theory in the 1960s and in reality, at least embryonically, in 1969, it wasn't until the early 1990s that that this technology was to "take off" with the run-of-the-mill American population. The reason for this rapid adoption is due to a number of different factors, but certainly the appearance of user-friendly graphical user interfaces (GUIs) to "browse" or "surf" the Internet was one of the most critical factors. Using Web browsers that allowed a person to "click and point" his or her way around this increasingly diverse and global network content demystified the interactive computer experience for many people, although, as just mentioned, this was only one of many factors. Digitization, expanding computer networks, greater bandwidth technologies, federal initiatives, educational programs, and a receptive market, as well as other reasons, contributed to the growth of online industries. As is still the case today, technologies were converging. It was becoming possible to send not only text but pictures, sound, and video over telephone lines. Newspapers and telephone companies were working together to offer online services. Television and software companies were looking at delivering content on computer screens as well as television screens. The borders that had separated communication industries and their content and services were beginning to blur.

It seemed inevitable to federal lawmakers that a massive rewrite of the Federal Communications Act of 1934 was on the horizon. After years of negotiation and discussion, and versions of the bill passed in the U.S. Senate and House of Representatives, President Bill Clinton signed the Telecommunications Act of 1996 on February 8, 1996. Judging from the rousing response from the telecommunications industry as a whole, the law seemed to please the private sector as it signaled a future of decreased government regulation and open competition in the telecommunications sector.

Federal law of this scope and complexity is difficult to summarize in adequate detail (the interested reader should seek out more comprehensive sources) but key points of the law include the following:

- There are fewer regulations on media ownership. Ownership caps on the number of radio and television stations a single company can own in a single media market, for example, is liberalized.
- Cable television benefits from fewer regulations. Cable companies have more freedom to set their own rates and engage in other services outside of traditional cable television, such as telephone service.
- With conditions, long-distance telephone companies can offer regional service, and regional Bell operating companies can offer long-distance service.
- The FCC has exclusive jurisdiction to regulate the provision of direct-to-home satellite services. This pertains to the direct broadcast satellite (DBS) market that has been growing in popularity in recent years.
- Generally, the FCC should reduce regulations to encourage more competition in the telecommunications sector.

There were also terms related to the regulation of controversial content. The installation of V-chips in newly manufactured television sets would give parents the discretion of blocking unwanted content (e.g., violent and sexual content), and a proposed television ratings system would work in conjunction with such technology. More problematic, however, was a portion of the law called the Communications Decency Act (CDA), which was meant to censor not only obscene but also indecent content. (Obscene content is not protected by the First Amendment, but indecent content is.) The CDA was challenged by free speech advocates and, as expected, was struck down by the U.S. Supreme Court in 1997. The Supreme Court affirmed a lower-court ruling that said the CDA violated the First Amendment. The American Civil Liberties Union (ACLU) and the American Library Association (ALA), both of whom led the fight against the CDA, argued that the law was too broad, was too vague, and infringed on free speech rights. The Supreme Court's decision was a victory for free speech advocates as well as for Internet companies, libraries, online publishers, and others who might have been negatively and unfairly affected by the broad and imprecise restrictions.

But the CDA was just the beginning of attempts to grapple with the slippery regulatory slope of digital communication. In 1999, Congress passed the Child Online Protection Act (COPA), which, like the CDA, has since been challenged by the ACLU on constitutional grounds. Rules and regulations in cyberspace will continue to evolve as this digital environment matures. Concerns over intellectual property, privacy, obscenity, piracy, fraud, search and seizure, trespass, censorship, and other issues that extend into the digital world have found their way into courts and legislatures. The balance between free speech and government regulation continues to be tenuous, especially where the protection of children is concerned. News stories of electronic trespass and criminal destruction or mischief are appearing more and more often.

The controversial Digital Millennium Copyright Act,[2] enacted in 1998, revised and reinforced existing copyright laws in light of a digital media environment. The law provided legal remedies against those who willfully use technologies to "circumvent" copyright protection systems. It implemented a 1996 World Intellectual Property Organization treaty designed to get signatories from different nations to help enforce each other's intellectual property. This law was supported by commercial content producers, such as software makers and record companies, whose digital products are prone to piracy because of the ease of copying and distribution. But it was a concern to many academics, librarians, and libertarians who thought it might be an impediment to fair use, the clause in intellectual property law that allows the copy and distribution of copyrighted material under certain conditions. The conventional wisdom surrounding the purpose of the DMCA, however, was that it would discourage the illicit copying and distribution of intellectual property on the Web, which was becoming rampant on a global level.

On the financial front, much attention has been paid to issues of domestic taxes and tariffs. In 1999, the World Trade Organization (WTO) agreed to shelve discussion of e-commerce issues prior to its disastrous meeting in Seattle, Washington, and extended a 1998 agreement by trade ministers that put into effect a moratorium on customs duties of electronic transmissions. However, the organization resurrected discussions in the summer of 2000.

In the United States, the federal and state governments are trying to iron out tax policies that apply to e-commerce. In September 2000, the California state assembly proposed e-tax legislation that would require local retailers to charge customers a sales tax on goods sold through the Internet. Other states are looking into some kind of taxation structure as well as a way of increasing existing tax revenue streams. As it stands, if a customer buys a book from a traditional bookstore in a state with sales tax, that customer will have to pay a sales tax on the book. But if he or she decided to buy that book online from the bookstore's Web site, there would probably not be a state sales tax added. Local off-line retailers tend to support an Internet sales tax because Internet sales could present unfair competition otherwise. The retail merchants lobby has been trying to persuade Congress not to allow cyberspace to remain a tax-free zone. States fear that they will increasingly lose tax revenue to the tune of hundreds of millions of dollars a year if sales are conducted over the Internet as opposed to from traditional points of sale, unless Internet taxes are imposed. Some estimate that number will rise to the billions in a few years, which could negatively impact local social services, education, and other programs. But so far, monitoring and collecting e-commerce transactions are at best difficult and certainly unpopular with the public.

The federal government has agreed to extend a moratorium (the Internet Tax Freedom Act of 1998) on new Internet taxes until 2008. Politicians who oppose new Internet taxes believe it will prematurely hamper the growth of a potentially prolific "place" for robust trade and commerce. Antitax supporters also argue that online retail sales are currently so miniscule (less than 1 percent) of all retails sales combined that it is too early to be worrying about the effects of not taxing Internet sales. The task of trying to tax customers from different states a different sales tax and then paying these taxes to the respective states understandably strikes the e-commerce community as daunting and not currently cost effective.

Free Speech versus Copyright Infringement

One of the most contentious debates concerning the regulation of cyberspace (aside from online pornography) has revolved around the violation of intellectual property. It is no exaggeration to say that copyright infringement on the Internet is rampant, and the action is often passed off as falling under free speech protection. Copyrighted articles, photos, music, screenplays, poems, drawings, and other intellectual property are routinely posted on Web sites or e-mailed without the copyright owner's permission. Intellectual property law, however, clearly protects the author (which, in this case, includes the photographer, artist, lyricist, playwright, etc.; in other words, the owner of the copyright)[3] and gives him or her the sole right to duplicate the copyrighted work, which must be an original work in fixed form, such as a book, CD, magazine, and so forth. Violations of copyright are actionable in a court of law and can result in serious financial penalties. There are four major types of intellectual property law in the United States: copyright law, trademark law, patent law, and trade secret law. Copyright is of primary interest here.

In addition to having the exclusive right to reproduce their own work, authors also have the right to modify their works and create derivative works based on their original work; distribute copies of their work to the public by sale, rental, lease, or lending; publicly perform their work; and publicly display their work. The Internet makes it easy to infringe on someone else's copyright. Digital scanners, e-mail, the Web, file transfer protocol (FTP), and other technologies are often employed in the act (whether intentional or unintentional) of copyright infringement.

The best way to avoid violating copyright law is not to copy an author's original work without permission. Getting permission from a copyright owner to use a work is obviously the most logical step toward observing copyright law.

Two high-profile cases in the news in 2000 involved allegations of copyright infringement by the Internet music-sharing companies Napster and MP3.com. Other companies offer similar services. Using innovative file-sharing software, millions of Web users, many of them college students, could share and download "free" CD music as MP3 files.[4] Both companies became embroiled in serious legal entanglements with some of the most powerful forces in the music industry. Universal Music Group, the world's largest record company, alleged that MP3.com failed to get permission from the copyright owners of the music that was collected and digitally stored in MP3.com's popular music database, which was publicly accessible to Web users. On June 9, 2000, MP3.com agreed to a settlement with other record industry labels—Sony Corp., Time Warner Inc., EMI Group, and Bertelsmann—after a U.S. District Court judge ruled in April that MP3.com had violated federal copyright law. Universal chose to press on with its lawsuit. In September 2000, a federal judge found MP3.com guilty of violating Universal's copyrights. Damages could reach as high as $250 million pending appeal. Frightened by legal ramifications, many colleges and universities have blocked access to Napster. Others resort to warning their students about potential copyright infringement using these kinds of technologies. Certainly, there are at least two minds about MP3 file swapping, and they couldn't be further apart. Some view the act as liberating music from the greedy and gluttonous control of record labels; others see it as stealing.

As the Napster and MP3.com cases suggest, the rules and regulations that pertain to digital communication and cyberspace are still in formation. As problems arise, governments and regulatory bodies will need to respond with legal and policy remedies at both the national and international levels. Some think a new body of "cyberlaw" or "netlaw" will have to be developed. This may be practicable (perhaps!) within the confines of national boundaries, but cyberspace transcends well-contained geophysical borders. The question remains: Are there mechanisms that can deal with developing rules and regulations in cyberspace?

International Regulations

As the Internet and World Wide Web and other communications systems become more entrenched as interactive global media, the question of international regulation constantly arises. Who, for example, is responsible for monitoring, investigating, and penalizing criminal behavior over the Internet when people from many different nation-states are involved? Does each country's own law enforcement agency step in and try to work together cooperatively? Does an existing international law enforcement organization such as Interpol have jurisdiction? And what if an entire country is suspected of foul play over the Internet, threatening the security of another country's communications network through covert cyberterrorism or other forms of premeditated digital attack? What is the appropriate venue for bringing such a case to trial?

The answers to these questions probably lie in history. How has the world, since the era of modern telecommunications began in the 1840s, managed (and fairly successfully at that) to achieve international cooperation for its increasingly global and interdependent telecommunications infrastructure? The answer is intergovernmental organizations. The United Nations already oversees well over a dozen agencies charged specifically with communications issues. Most people probably never give much thought to many of these organizations, but without them, our airspace and oceans and mail service would be chaos. Air travel is safer because of the International Civil Aviation Organization (ICAO). Transportation over the seas has benefited from the International Maritime Organization (IMO). Mail delivery runs more or less smoothly between nations thanks to the Universal Postal Union (UPU). Our radio signals don't clash because of negotiations and assignments in the International Frequency Registration Board (IFRB) and the International Radio Consultative Committee (CCIR). The International Court of Justice attempts to uphold international law, as difficult as a task as that might be. The list goes on. Messy acronyms, yes, but extraordinarily important, every one of them.

Two intergovernmental organizations may be pivotal in drafting global communication policy: the United Nations, which has been helping formulate international agreements for more than 50 years, and the World Trade Organization (WTO), which is emerging as the authoritative body in matters of global trade and commerce. It's worth pondering what role these two omnipresent institutions might play in helping develop rules and regulations in global communication. Like the International Telecommunications Union before them, the UN and the WTO could well be the best forums for planning and coordinating the global telecommunications infrastructure of the next millennium. They are, of course, not without controversies and problems, as any institution of their size and scope are bound to have, and whether they can adequately represent the diverse cross-section of nations and interests that are a part of the global community is an open question. It seems likely that national governments and intergovernmental organizations, such as the UN, its various agencies, and the WTO, will need to collaborate on how to best regulate the burgeoning global telecommunications network and its content. These organizations must also be sensitive to the global inequities surrounding media and information technology. The special needs of developing countries need to be taken into consideration when embarking on international communication policy and planning. The UN has tried to do this, sometimes to the chagrin of the United States, but the WTO is less trusted by many grass-roots organizations because of perceptions that it operates too often in secret and under the aegis of First World countries.

Much activity has been occurring on a national level, as well, to help combat cybercrime globally. Former U.S. Attorney General Janet Reno created division of the Department of Justice to focus on cybercrime. In her February 16, 2000, testimony to members of the U.S. Senate serving on various relevant committees, she outlined the need for her office "to combat the growing problem of cybercrime, particularly in light of the recent denial-of-service attacks on several major Internet sites." She called for a long-term strategy to deal with cybercrime and highlighted the following things she would like to see done to adequately deal with the problem of cybercrime:

- Developing a round-the-clock network of federal, state and local law enforcement officials with expertise in, and responsibility for, investigating and prosecuting cybercrime.
- Developing and sharing expertise—personnel and equipment—among federal, state and local law enforcement agencies.
- Dramatically increasing our computer forensic capabilities, which are so essential in computer crime investigations—both hacking cases and cases where computers are used to facilitate other crimes, including drug trafficking, terrorism, and child pornography.
- Reviewing whether we have adequate legal tools to locate, identify, and prosecute cybercriminals. In particular, we need to explore new and more robust procedural tools to allow state authorities to more easily gather evidence located outside their jurisdictions. We also need to explore whether we have adequate tools at the federal level to effectively investigate cybercrime.
- Because of the borderless nature of the Internet, we need to develop effective partnerships with other nations to encourage them to enact laws that adequately address cybercrime and to provide assistance in cybercrime investigations. A balanced international strategy for combating cybercrime should be at the top of our national security agenda.
- We need to work in partnership with industry to address cybercrime and security. This should not be a top-down approach through excessive government regulation or mandates. Rather, we need a true partnership, where we can discuss challenges and develop effective solutions that do not pose a threat to individual privacy.
- And we need to teach our young people about the responsible use of the Internet.[5]

The problem of cybercrime is more critical now than ever. After the terrorist attacks on the United States, concern about cyberterrorism heightened considerably. Even before those attacks, however, government, academic, and corporate computer systems were routinely broken into by hackers, and viruses of one kind or another have been purposefully spread over the Internet from both within and outside U.S. borders.

Cybercrimes can take many different forms: harassing or stalking online, stealing private information from a secure computer, transferring money illegally electronically, selling drugs or other contraband through the Internet, soliciting illegal activities such as an adult seeking sex with a minor or someone seeking a partner in crime through e-mail, software piracy, and so forth. What makes the problem so vexing for law enforcement personnel is the difficult nature of tracking down violators and the lack of coordination among law enforcement agencies across national borders. That seems to be changing with more international agreements vowing to cooperate with each other to crack down on cybercrime.

For example, in August 2001, U.S. Attorney General John Ashcroft announced that 100 people had been arrested in relation to a commercial child pornography site on the Internet. Although a U.S. couple was prosecuted for distribution of the material, many of the people who provided photos and video of child pornography lived outside the United States, as did others who paid to access those images. According to news reports, the site had more than 300,000 subscribers from around the world. One can imagine the difficulty a government would have trying to bring criminal charges against the citizens of another country, especially when laws and penalties pertaining to certain behaviors are not consistent from country to country.

As part of a larger antiterrorism package, the U.S. government is investing heavily in antiterrorism surveillance and monitoring technologies and has put Americans on the alert for possible terrorist activity online. Such activity can be both national and transnational in scope. A coordination of law enforcement cooperation throughout the world regarding cybercrime is just one of many challenges that make regulating the Internet beyond national boundaries so challenging. If the long history of international communications policy and planning is any indication, we can bet that the exponentially more complex global regulatory road ahead will be a long and rocky one.

Questions for Discussion and Comprehension

1. What is the purpose of laws and regulation in a society? As the Internet becomes integrated into society, what are ways of creating rules governing this "sphere of interaction" that take into consideration the public's right to communicate and the government's responsibility to insure public safety and order?
2. How should commercial transactions on the Internet be taxed? Research current tax structures regarding commerce in the real world, and then suggest a way that online transactions should be taxed. Or argue for a nontaxed environment.
3. As global interconnectivity increases, the possibility of global crimes increases—for example, the trafficking of obscene or other illegal content. What are the complexities of creating laws to enforce this trafficking? What are possible solutions to try to curb global trafficking of illegal content?

Endnotes

1. The full text of "Areopagitica" can be found on a number of different Web sites, including the Milton Reading Room at www.dartmouth.edu/research/milton/reading_room/ and search for Areopagitica.

2. H. R. 2281, The Digital Millennium Copyright Act, was signed into law by President Bill Clinton on October 28, 1998.

3. The author may not always be the copyright owner. If the work was created within the scope of the author's employment, the author's employer may be the copyright owner. Or work may have been created as a specially commissioned work for hire, in which case the commissioning entity may enter into an agreement of copyright ownership with the author prior to commencement of the work.

4. MP3, or MPEG-1 Audio Layer 3, is a format popular for storing music files. It significantly compresses the music file without degrading the sound quality.

5. Statement of Janet Reno (then) Attorney General of the United States before the United States Senate Committee on Appropriations, Subcommittee on Commerce, Justice and State, the Judiciary and related agencies, February 16, 2000.

8

News Media and Society

Role of the News Media

The news media in the United States, and even during colonial times, have long served as an important and increasingly tenacious check on the power and influence of government and big business. A number of early newspapers in the American Colonies, for example, represented the brains and brawn of their fiercely independent editors as they attacked displays of arbitrary power and relentlessly challenged an oppressive status quo, often to society's long-term benefit but to their own personal detriment. Editors James Franklin (the older brother of Benjamin), John Peter Zenger, and others served time in jail for espousing views that were distasteful to the British Crown and its appointed representatives in the Colonies. In a later era, muckraking journalists, such as Ida Tarbell and Lincoln Steffens, wrote investigative articles exposing corporate and government corruption. Although no longer called muckraking, investigative journalism is still alive and well today and has been instrumental in advancing honesty and fairness in government.

These provocative voices set the stage for a society in which the mass media—at first newspapers but later magazines, radio, television, and now the Internet—could serve as agents of social change by introducing ideas and information, sometimes regarded as dangerous and revolutionary by society's power brokers, into the general population without fear of retaliation (outside of legal means) by those who were being reported on. The news media also call attention to problems around which some kind of social mobilization is already occurring. By focusing their attention on these problems, they also contribute to focusing public's attention on these problems as well, which can help to flame the fires of social activism.

The news media are a diverse lot, however, and to paint them with one broad brush stroke would be distorted and inaccurate if this were a book about news. Suffice it to say that although tremendous differences exist from medium to medium and between particular news organizations within each medium, there are still some generalities that can be made. The news media, while themselves businesses and increasingly concerned with profits, still function primarily as a public service through the work of their journalists and editors who gather, analyze, and communicate news of current interest to respective media audiences.

Theoretically, journalists are not beholden to the influences of government and the marketplace but rather strive to report the "truth," a practice grounded in a tradition close to the heart of American democracy. Throughout history, such reporting has often pitted the news media against government and corporate interests but, at the same time, has won the journalism profession a place of special distinction in society as a trusted observer, reporter, and watchdog working on behalf of the public. Although this trust may occasionally be violated, the public expects that professional journalists and their respective organizations will hold themselves to a higher standard than those who do not work in the public interest.

Perhaps the most notable modern example of the "power of the press" occurred in the early 1970s as the notorious Watergate scandal unfolded. The 1974 resignation of President Richard Nixon might have never occurred without relentless and tenacious investigative reporting by the nation's news media, in particular the *Washington Post* newspaper. This is one of thousands of importance cases in which the investigative prowess of individuals in the news media has led to important social reforms. Almost every serious news organization can boast of stories that have helped improve their community in significant ways by alerting the public to issues or problems that needed their attention and action. In many instances, the news media have been the only institution influential and pervasive enough to expose large-scale criminal or unethical practices, resulting in corrective action in response to public pressure.

Social critics do not always see the role of the news media as ideally as has been presented here. Some see the news media, and all commercial media, as perpetuating dominant ideologies and limiting the scope of public discourse through a process of gatekeeping, agenda setting, and framing. A large body of literature supports this view of the media, and much of it is thought-provoking and philosophically rich. This chapter takes a more traditional view of the news media, however, as watchdog and vehicles for free speech and other democratic ideals.

Because the news media can generate not only public awareness about problems and injustices in society but also mobilize large-scale human response of one kind or another, the news media are thought to wield an influence comparable to society's most powerful social institutions. It is this influence that both benefits and can be marshaled by the forces of civil society—an abstract concept that represents those in society who are not primarily motivated by their official attachment to government or business but who care about bettering some aspect of their community (as they define it) outside of formal state or corporate structures. The balancing act among this array of social forces (i.e., government, business, and civil society) may often seem to be tenuous, but they each have sufficient influence in different ways (e.g., government through laws and public policy, market through capital and labor, and civil society through public opinion and mass mobilization)[1] to keep a check on each other's relative power in society. Digital media technologies have become a particularly useful tool for civil society.

Civil Society and Digital Media

The concept of a *civil society* is one muddied by multiple definitions and different levels of complexity. Its roots are equally disputable. Many scholars typically look at eighteenth-century European society as a logical cradle of civil society—a time when "the centralized authority of Rome placed power in several sets of hands" (Hall, 1995, p. 4). Some of the key ideals of the Enlightenment—espoused by intellectuals who believed that one's inalienable rights, combined with the power of reason, could be used to combat tyranny and work toward a more just society—provided a launching pad for the emergence of civil society. Along with the rise of literacy, a middle class, and the means to disseminate ideas through mass printing technology came the emergence of a group of people who deliberated over such topics as culture, art, and politics. Over time, their ideas were important in counterbalancing the hitherto overwhelming influence of church and monarchy. Civil society was born of this socially mannered yet radical intellectual movement. Today, the term is applied to many different instances where voluntary associations of people work toward peace, social justice, environmental protection, civil rights, and other important causes.

For the purposes of this chapter, the term *civil society* is simplified to describe that rather amorphous sector of society that is not government and not business but a nonprofit or voluntary association of people working together toward a common goal of social improvement. The goal could be related to the environment, population, human rights, health care, equality, diversity, peace, disarmament, free speech, Third World development projects, international relations, homelessness, poverty, and a host of other issues. People, motivated by a sense of shared social responsibility and not political power or corporate ambition, organize around a problem and work collectively toward a solution.

In small, geographically contained locations—a small town, a village, a rural community—this organization and mobilization can take place through traditional nonmediated channels. The individual is the agent of social change; face-to-face communication is the way that information and persuasion is exchanged from person to person. In a small town, people can mobilize against the construction of a parking lot, or in favor of extending a schoolhouse, or in support of a family visited by tragedy. Meetings can take place in the town hall, a church building, or other geo-physical commons; people can deliberate among their appointed leaders; decisions can be made, postponed, or reversed. Civil society at this scale is relatively simple and "direct," unencumbered by mediating technologies and interfaces.

As populations grow and disperse, and problems and issues are of a much larger scale than those that confront a village or small town, the means that civil society uses to communicate extends beyond the face-to-face channels into various interpersonal channels. These include mass media and, more recently, digital media—electronic mail, electronic listservs, Web pages, chat rooms, and so forth.

The traditional mass media can facilitate the goals of civil society (or can ignore them or, in effect, sabotage them), depending on the nature or even the existence of the media coverage. A grass-roots movement to save an endangered species, for example, can receive mass media coverage that alerts others to the problem and increases public awareness and public deliberation, perhaps even resulting in a discernible level of public opinion that influences decision makers to support the goals of the movement. On the other hand, the mass media could ignore the movement altogether, or frame the problem in a way that is not helpful to the movement (e.g., trivializing its importance or raising counterclaims that challenge the veracity of the movement's claims). In any event, civil society can *try* to marshal the forces of the mass media to advance the goals of a particular movement or cause (and it makes sense to try, since the mass media can be an influential social force). But it cannot dictate the nature of that coverage and thus cannot really predict whether the coverage, if any, will be helpful, harmful, or ineffectual.

In eighteenth-century England and France, intellectuals would sit in coffee houses and "salons" and give their opinions about a wide range of topics that concerned them about society at the time: literature, art, culture, politics, religion, philosophy, and so forth. Today, those public spheres still exist, but they've also been extended to a technologically mediated conceptual space that allows conversation and public deliberation to occur among people who cannot come together physically and, in fact, may never have met each other face to face. In a digital media environment, the members of civil society can communicate, organize, and mobilize via more user-driven communication technologies, such as the Internet and World Wide Web. Civil society, some would argue, has been evolving in cyberspace as scholars, activists, volunteers, nongovernmental organizations, nonprofit organizations, environmentalists, cults, vegetarians, nationalists, and even racist and anarchists have used the Internet to communicate, organize, and mobilize toward common objectives. Just about any cause, no matter how absurd or misguided, can probably find a sympathetic audience among at least a handful of people on the Internet.

Increasingly, significant social events coevolve on the Internet and in real life (or in the "real world"). An example that has been highlighted in recent times has been the massive protest against the World Trade Organization (WTO) that occurred in Seattle in late November and early December 1999. Activists throughout the world had reportedly been discussing and planning WTO protests online for about a year prior to the actual event. According to an article in the *Los Angeles Times*, e-mail began "ricocheting" from one part of the globe to another, alerting anti-WTO activists about plans to converge on Seattle for a massive protest of WTO. Eventually, a variety of digital media were employed to get the word out about the now infamous "battle in Seattle." Digital media technologies employed were listservs (structured e-mail mailing lists), Web sites, and other electronic communication tools. "These were the digital origins of what has become one of the most incendiary U.S. political uprisings in a generation," the article said.[2]

The Internet provided a quick, cheap global linking device for a diverse group of people and organizations to discuss the WTO's then-upcoming ministerial meeting in Seattle and collectively chart a course of action against it. Coalitions and alliances were formed, a Web site was formed to facilitate global information sharing among these groups, and residents in Seattle who had lodging available were linked up with protestors arriving from outside the city. The *Los Angeles Times* article said that many protestors relied exclusively on online information to prepare them for their journey to Seattle.

One protestor interviewed said that years ago, her organization got hold of a "sensitive document" relating to the WTO, took the document to a commercial copier for duplication, and mailed the copies via overnight express to 30 recipients. Now they get documents, scan them into a computer, and disseminate the information all around the world at lightning speed.

The WTO protests in Seattle are a high-profile example of online organization and mobilization at work, but many more examples can be invoked that are far less attention grabbing. In fact, such activities occur on a daily basis and result in constructive albeit less visible outcomes. The Institute for Global Communications Web site, for example, calls itself the "gateway to the progressive community online." The site has links to Jubilee 2000 (the group advocating Third World debt relief) and to Web sites devoted to topics such as peace, the environment, voting, women's issues, and anti-racism.

Setting up a Web site has become easier and more affordable than ever. Of course, the traditional mass media are still important in the communication activities of civil society. When the news media cover issues that a particular group of people considers to be important, for example, that media attention can shed light on a social problem that might otherwise go unnoticed. In addition, audience members (of traditional news media) may decide to get involved with a social problem or event in some way (e.g., giving money, writing letters, attending a demonstration, boycotting advertisers, volunteering, etc.). But civil society is no longer as dependent on the traditional news media as it once was because information sharing and social organization can also take place on the more versatile and accessible Internet.

The public demonstrations following the controversial presidential elections in the United States is a good example of how the Internet was used as a tool for social mobilization, allowing people to bypass the traditional media and communicate with each other through interactive electronic channels instead. After it was revealed that voting "irregularities" may have occurred in the state of Florida during the general election, affecting the state's popular vote for president, a national movement got underway to demand a recount and revote. Organizers of public protests in cities all over the United States used the Internet to air their concerns about the elections and to organize nationwide protests.

One Web site, calling itself CounterCoup.org, featured a list of nearly 100 cities and gave specific locations where protests were scheduled to be held on the two weekends following the general election. The site also featured links to electronic message boards for specific cities to help local protest organizers communicate with each other in their particular geographical regions.

Less than 24 hours after the first demonstrations occurred throughout the nation, dozens of attendees from different cities had already posted their impressions of the event on the CounterCoup Web site, with some even posting photos. Turnouts were varied. One report from Las Vegas read, "We rallied today in Las Vegas in front of the Federal Bldg... Only about 12 of us showed up, but we got on the evening news... three TV stations came out... no newspapers, but we got our point across."[3]

An account from Los Angeles painted a different picture:

> "LA rocks! There were at least 1,500 people at the Federal Bldg. at 11000 Wilshire when I left at 3 p.m. All four corners of this busy intersection in Westwood were packed and the crosswalks full every time the lights changed. Lots of sign-making materials on hand—tons of signs and enthusiasm! Huge response in favor from passing cars! Honking, thumbs-up, shouts—lots of excitement! An impromptu percussion band played on the NW corner. Many camcorders and cameras on hand recording it all. I saw the local ABC affiliate and lots of press photographers. Fantastic![4]

In Seattle, more than 100 protesters gathered in front of Seattle Central Community College in the Capitol Hill district to voice their grievances and cheer on speakers from the crowd who had access to a hand-held bullhorn. The *Seattle Post-Intelligencer* reported that the idea for the demonstration in Seattle spread through the Countercoup Web site.

Other "protest" Web sites encouraged visitors to write to government officials and express their opinions about the current situation in Florida. A site called WorkingForChange.com, sponsored by the Working Assets telephone company, allowed visitors to fill out an online form to send an e-mail to Florida's then-Secretary of State Katherine Harris, urging her to push for a revote. A prewritten message could be used for the e-mail or an original message could be written in its place.[5] A common practice on activist sites, users of this site could also look up directory information about other elected officials and send them e-mail about this or any other issue through the WorkingForChange Web site.

A Web site called Trust the People! took yet a different approach, appealing directly to Palm Beach County, Florida, voters. The site, which has ties to a Democratic organization, provided Palm Beach County voters with a printable "affidavit you will need to officially participate in a challenge [of] the results" and explains: "If you voted in Palm Beach County and believe your vote was improperly recorded you must complete an affidavit as specified in Florida election law. The affidavit contains a fax number of a local law firm that is participating in the collection and filing of these affidavits."[6]

As was also the case during the months leading up to the World Trade Organization demonstrations in Seattle in late November and early December 1999, the Internet seems to emerging as the communication tool of choice for organizing mass protests around controversial issues.

Digital Empowerment

One of the promises of digital media in the mid-1990s was customization, the ability for individual users to specify the news that they wanted, when they wanted it, and in the amount that they wanted it. The user-driven technology of the Internet and World Wide Web was supposed to transform the "push" model of mass communication to the "pull" model of personalized information seeking.

The concern with the "pull" model, however, was that people might only seek information that they want and not have exposure to a well-balanced menu of news and information. Some critics felt that this could lead to a society in which people become increasingly ill-informed about social and political issues and thus would not be as civically engaged or able to make responsible decisions as an electorate.

Those concerns are still valid and should not be abandoned, but it appears that our worst fears have not been realized; in fact, online news and information have been additive and not substitutive when it comes to information seeking practices for a large number of users. Studies have suggested that people who use the Internet for getting news are also heavy consumers of traditional news media sources. The Internet allows people to get more, not less, news than they were previously getting from exclusively traditional media sources. If some users do become excessively narrow information-seekers, accessing mainly sports or entertainment news all day everyday, for example, to the detriment of other types of news, it is doubtful that they would be substantially more conscientious consumers of general news if the Internet did not exist.

The Internet does not seem to be responsible for creating a society in which the population is less informed or getting a less balanced diet of news and information. If people are not as intelligently able to participate in civic activities today than a couple of decades ago, the problem began long before the 1990s and is much more complex than any single possible causation. One could persuasively argue that the Internet might actually facilitate civic engagement for those who otherwise might choose not to get involved with public affairs. The large number of protest movements, for example, that use the Internet as a communication medium to share information and help mobilize people from all parts of the country or world suggest that the global environmental, labor, human rights, health, and other social movements actually benefit from the ease of peer networking that the Internet offers.

Some scholars, such as Howard Frederick, have even suggested that the Internet has facilitated the formation of a "global civil society." These kinds of global civic networks were much more difficult to create prior to the Internet. Today, large-scale social action among people from many different geographic locations can benefit from the ease of horizontal (as in more direct and participatory) communication rather than the vertical, hierarchical, limited interaction model of traditional mass communication. "People power" at all levels depends on having open channels of communication among, well, the people!

Although it may often seem like it in the news media, empowered groups do not make their collective feelings known only at mass demonstrations. These groups find expression through voting, consumer decisions (e.g., boycotting advertisers), the referendum and initiative process, public discourse, contributions (financial and otherwise) to political action groups, volunteerism, and so forth. History has shown that "people power" is a force to contend with in any society—indeed, entire governments have come crumbling down after public opinion has turned against them, even those under dictatorships.

In a democracy, voting is arguably the most powerful way the public can express and impose its will, because a society's top lawmakers are put at the will of the people. According to the U.S. Census Bureau, however, a large number of Americans do not avail themselves of this right. A 1998 Census Bureau report said that only 42 percent of the 1998 voting-age population reported voting in the 1998 congressional election, the lowest percentage since the Census Bureau began collecting such statistics.[7] Why is this? Low voter turnout, some argue, reflects an increasingly cynical public whose trust in government is on the wane. It could also be that as society grows larger and less personal, the opportunities for public deliberation decrease, and the sense of civic engagement is weakened. There is less interest in the democratic process since one is less involved in politics, and voting becomes more of a spiritless submission to social obligation rather than a passionate exercise of social responsibility. It is easy not to conceive of one's role in civil society without a sense of belonging to a community or without a belief that political action on the part of an individual in a group will make any difference. Relatedly, the larger a society grows, people on the margins of social power may feel invisible and inconsequential.

Digital media are not a panacea for strengthening the democratic process and reversing voter apathy, but it could play a significant role in rebuilding a sense of civil society and public deliberation. The technology in and of itself will not make any difference in creating increased civic engagement, but the *application* of the technology by those who believe it can be a tool for more democratic communication and action is what holds promise. Human initiative and the collective will of a critical mass of people to exploit digital media for civic engagement are essential if civil society is to derive social benefits from digital media the way that the private sector hopes to derive considerable financial benefits.

Questions for Discussion and Comprehension

1. What role in history have communication media, such as pamphlets, magazines, newspapers, radio, and television, played in building a democracy and serving as a watchdog of government? Can the Internet play a similar role, and if so, how?
2. Discuss one way that you could use the Web as a tool for political protest or consumer education. These are uses that do not have profit making as an ostensible goal. Suggest a real-life example of how the Web can be employed as a tool for prosocial action by civil society.
3. What do you think of Internet voting? Good idea? Bad idea? Explain your position.

Endnotes

1. Of course, the basis of power for each of these sectors overlap. Government, for example, has wealth of its own as well as its own labor force. These characteristics are meant to be suggestive of dominant (rather than exclusive) influences.

2. Greg Miller, "WTO Summit: Protest in Seattle," *Los Angeles Times*, December 2, 1999, p. A-24.

3. From the CounterCoup Web site, http://www.geocities.com/countercoup, November 12, 2000. This Web site is no longer updated on a daily basis.

4. Ibid.

5. From the WorkingForChange Web site at http://www.workingforchange.com, November 12, 2000.

6. From the Trust in the People! Web site at http://www.trustthepeople.com, November 12, 2000.

7. U.S. Census Bureau, "Voting and Registration in the Election of 1998," *Current Population Reports*, August 2000. It should be noted that historically voting and registration rates are lower during nonpresidential election years. These rates are compared with nonpresidential election years.

9

Research and Inquiry

Mass Communication Research in General

Studying digital media in academic circles can take many forms. Engineers would likely focus on software and hardware development, and, to a certain extent, on usability issues (e.g, how easy or difficult are the technologies to use, and what could make them better?). Library and information specialists might focus on data archiving, search mechanisms, user-interfaces, and electronic research resources and techniques. Legal scholars would examine the changing laws and policies resulting from the emerging digital media environment, including complex issues related to intellectual property, privacy, cybercrime, and indecency or obscentity.

In the social sciences, the effects of digital media—the technologies themselves, their content, and the communication process—are of considerable concern. This is not surprising, because the social sciences have long been concerned with the effects of media on society. Even as far back as the late 1800s and early 1900s, sociologists were discussing the role of newspapers in forming a kind of nonphysical community. In 1901, the French sociologist Gabriel Tarde (1969, p. 277), for example, tried to describe the psychology of the public—"a dispersion of individuals who are physically separated and whose cohesion is entirely mental"—that arose when geographically dispersed people formed a public opinion about a particular issue or problem by reading their newspapers. Similarly, Robert Park in the early 1920s began studying the role that newspapers played in helping people in the cities—whose ties to each other were not strengthened by widespread interpersonal ties (as they were in the villages)—think of themselves as a single community.

The research of the process and effects of media continued throughout the decades. The Payne Fund Studies in the early 1930s researched the effects of movies on children. This was in response to the rapid growth of the film industry in the previous decade, which attracted throngs of enthusiastic viewers, including a large number of children. Social scientists were hired to conduct studies about the effects of movies on their audiences. One of these was a sociologist named Herbert Blumer, who eventually wrote a book called *Movies, Delinquency, and Crime* in 1933 based on his Payne Fund research. By today's standards, the methods used to link movies with juvenile delinquency are questionable, but the point is that this kind of research was addressing the public concerns in society at the time.

The history of media studies reveals decades of research concerning the effects of messages on audiences via a variety of media, such as newspapers, radio, and television. In the first half of the twentieth century, there was a growing body of research looking at the effects of propaganda, persuasion, and public opinion. World War II intensified this "effects" research tradition as governments waged propaganda and public information campaigns at their enemies and their own citizens alike. With every new technology come concerns about that technology's effects on society. The effect can be short term (temporary) or long term (persistent), and it can occur at different levels (cognitive, affective, or behavioral). In addition to the Payne Fund Studies, the effects of radio on audiences preoccupied a number of scholars. The famed Radio Research Project, started by Paul Lazarsfeld and later to become the Bureau of Applied Social Research at Columbia University, began by looking at the effects of radio messages on listeners. Early radio audience research involved studying the effects of radio soap operas, the drama "Invasion from Mars," and the sale of war bonds using a celebrity (singer Kate Smith) spokesperson on the radio.

When television became widely adopted by mass consumer society in the 1950s and 1960s, public concerns over that medium predictably increased. People wondered (and continue to wonder) about the relationship between aggression viewed on television and aggression exhibited in the real world. Is there a connection? The research findings vary.

Early research in social psychology by Albert Bandura (1977) suggests that children "model" behavior they see adults doing either in person or in some mediated form like TV or film. This proposition is known today as *social learning theory.* In his experiments, children were exposed to acts of aggression by adults, and then the children were mildly frustrated. Under certain conditions, children acted out aggressively as they had earlier witnessed adults doing. The experiments and explanations are more complicated than can be explained here, but this kind of research has been the basis for subsequent research that linked observed aggression or violence with expressed aggression or violence. One needs to keep in mind, however, that experiments conducted under laboratory settings do not necessarily tell us what will happen outside of the laboratory, where a host of other factors and influences come into play.

George Gerbner's Cultivation Theory[1] similarly links observed images on television or movies with perceptual changes in the real world. Gerbner argues that for a certain group of people, perceptions of reality are altered because of what they watch on television. These people, especially those who are heavy viewers of television and have fewer social outlets than the average person, are likely to have an unrealistic view of the world, thinking, for example, that it is a much more dangerous place than it actually is because of the amount of violence that they watch on television. Gerbner refers to this phenomenon as the "Mean World Syndrome." He also believes that regular exposure to the kind of entertaining violence that exists on many television shows and movies can lead to other negative effects, such as desensitization toward acts of violence or cruelty, a stereotypical view of both victims and criminals in society, and a need for increasing amounts of mediated violence for the same level of gratification to occur over time.

Effects Research and Digital Media

What kinds of effects research can be done on digital media? The body of effects research related to traditional mass media can help to answer this question. It is important not to look at media (including Internet) effects too simplistically. At one time, researchers may have thought that mass media had more powerful and undifferentiated effects than was actually the case. Metaphorically, we refer to such explanations of powerful effects as the Hypodermic Needle Theory or the Magic Bullet Theory, which implies that a message can be injected into (or aimed and shot at) an audience with predictable and planned consequences. In other words, one can change a mass audience's thoughts, attitudes, opinions, and behaviors by manipulating the messages they receive.

Today, few scholars would subscribe to such an uncomplicated view because of the totality of social science research that has been conducted over the decades. Clearly, certain media messages may have widespread impact, but audience members are not all affected in the same way. However, there may be types of audience members (e.g., those with baseline psychological pathologies) who may be affected similarly. But large audiences are composed of very different types of people—many are in relatively good mental and physical health, have a supportive external network (family, friends, colleagues, etc.), are of average or above-average intelligence, and are much more exposed to the world than just through the narrow view of a television screen or computer monitor.

Others who are less or differently endowed with the properties just mentioned may well react differently to media stimuli. This is an important point to keep in mind when doing any kind of media research. Even Gerbner makes important distinctions when claiming that violence on television has negative effects on audiences. Those effects may be worse for certain types of people, such as heavy viewers of television who do not have sufficient experiences and relationships in the real world to counteract the power of stereotypes.

Given this context, it is important when doing research about digital media that one's research questions are not too simplistic. Remember that different people are affected in different ways under different conditions by any kind of media, including Internet content or computer video games.

In the social sciences, a wide range of methods is used to investigate research problems. But before you embark on research, you should ask yourself these specific questions:

- What is your research problem? (A problem in this sense does not have negative connotations. It means, what specifically are you interested in doing research on? For example, your research problem may involve studying how a certain group of people use the Internet to find out information about personal health problems.)
- How would you formulate this into a problem statement?
- What are the terms in your problem statement that need defining?
- What method would you use to collect data about your problem?
- How would you go about collecting these data?
- How would you analyze these data?
- What else has been written about this research problem? What references or sources would you use to help you with this research?

There are many different methods that researchers use to answer their research questions. All of these methods cannot be explained thoroughly here (you need to consult a research methods book for a complete definition), but they are listed in case you want to learn more about them so that you can proceed with research on digital media in the future:

- *Survey Questionnaires:* A list of questions asked of individuals in writing, in person, or on the phone that are later tabulated and analyzed to study group responses.
- *Content Analysis:* A systematic study of artifacts, such as newspaper articles or television shows, to find thematic patterns that convey symbolic or ideological meaning to readers or audiences.
- *Experiments:* Carefully designed tests that are administered under controlled conditions so that comparisons can be made before and after a treatment or intervention is introduced.
- *Ethnographic Methods* (e.g., Participant-Observation): A field study of people and culture in natural settings.
- *Discourse* or *Textual Analysis:* A close examination of speech or writing for rhetorical elements and other patterns or themes.
- *Interviews:* Well thought-out and relevant questions asked of subjects who can help illuminate a research problem.
- *Case Study:* An intensive and comprehensive analysis of an organization, event, or person.

All of these methods can be used to do research on digital media. Say, for example, you wanted to learn about the computer-use habits of a group of high school students. You could come up with a set of questions (e.g., How many hours a day do you use a computer? Do you have access to a computer from home? Have you ever purchased anything online?) and ask these questions as part of a survey questionnaire. You might also go to a high school computer lab after school and observe what students are doing there and ask them questions. Much thought and preparation, not to mention expert guidance, goes into conducting good research. It is best to embark on any research project under the direction of an experienced researcher.

Digital Media Research

Digital media research is a new field that is interdisciplinary and dynamic. Scholars in psychology, for example, might look at problems of Internet addiction, obsessive-compulsive behavior, and relationship-formation online. Those in speech communications might examine online discourse and deliberation, communication in the absence of nonverbal cues, and the textual differences between e-mail and letter writing. Sociologists and political scientists could look at social structures in cyberspace, community building, and Web-mediated social action. Mass communication scholars might be interested in the integration of digital media services into traditional media organizations, the development and effectiveness of multimedia tools for telling news stories, and the changing role of journalists in cyberspace.

Many scholarly journals in the social sciences, humanities, and other areas are publishing articles that relate to digital media in some way. For example, the Spring 2000 issue of *Journalism and Mass Communication Quarterly*, featured an article by Guido H. Stempel III, Thomas Hargrove, and Joseph P. Bernt called "Relation of Growth of Use of the Internet to Changes in Media Use from 1995 to 1999." The method used for this research was a national survey, and the results showed that there was a large gain in the number of people who used the Internet during the period under study and a decline for local and network television news and for newspapers. The results further showed that "Internet users are more likely than non-users to be newspaper readers and radio news listeners. For both local and network television news viewing, there is no significant difference between users and non-users of the Internet. Clearly, the Internet is not the cause of the decline in use of the other media."[2]

The previous issue featured an article by Joseph R. Dominick called "Who Do You Think You Are? Personal Home Pages and Self-Presentation on the World Wide Web." This researcher studied personal home pages, which provided him with the opportunity to study the audience as producers of communication content rather than as consumers. After analyzing more than 300 personal home pages he found that most of them did not contain a lot of personal information. He found that the "same strategies of self-presentation were employed on personal pages with the same frequency as they were in the interpersonal setting. There were also gender differences in self-presentation that were consistent with research findings from social psychology."[3]

Of course, research of this nature is time sensitive. What may be true or apparent at one point in time may be different at another point in time. These are only two examples of the kinds of research people are doing related to digital media. A thorough review of the literature is important to get a sense for the larger body of research people have been doing over the years. But where does one look for this literature?

In addition to existing (traditional) scholarly journals publishing research on digital media, there have been new online and hard copy journals dedicated specifically to publishing digital media research. These include:

Journal of Computer-Mediated Communication
The Information Society Journal
New Media & Society
Information, Communication & Society
Convergence
First Monday
Electronic Journal of Communication
M/C—A Journal of Media and Culture
Journal of Electronic Publishing
Kairos: A Journal for Teachers of Writing in Webbed Environments
Communications of the ACM
Journal of Virtual Environments (JOVE)
Computer-Mediated Communication Magazine
Cybersociology
Journal of Online Behavior
Journal of Asynchronous Learning Networks
Journal of Interactive Media in Education
WebNet Journal—Internet Technologies, Applications & Issues
International Journal of Human-Computer Studies
Interacting with Computers
Computers in Human Behavior

These journals can be searched for on the Internet, or a reference librarian can help you locate them. This is not an exhaustive list, so there are others you should try to find. The first step to doing digital media research is to read as much as you can that's "out there." This helps you see what other researchers have done, how they have done it, and what they have found. Later, once you've determined what your research questions are, you will do a more formal and studied reading of the literature (a "lit review") pertaining specifically to your research. But in the initial stages, a broad exposure to the literature helps lay the scholarly foundations for future research.

A growing number of Web sites can help with Internet research. Here are just a few suggestions:

- *The Association for Internet Researchers:* Description from its Web site: "The Association of Internet Researchers is an academic association dedicated to the advancement of the cross-disciplinary field of Internet studies. It is a resource and support network promoting critical and scholarly Internet research independent from traditional disciplines and existing across academic borders. The association is international in scope." http://www.aoir.org
- *The Internet Studies Center:* Description from its Web site: "The Internet Studies Center at the University of Minnesota addresses compelling questions surrounding the social, ethical, legal, and rhetorical aspects of the Internet. The Center is based in the Rhetoric Department but is an interdisciplinary effort drawing on faculty from around the University and industry partners. By providing access to current theories, critiques, and explications of the Internet and its social aide, the Internet Studies Center is a hub for innovative Internet research and dialogue. Specific goals for the Center include: encouraging cutting-edge, interdisciplinary research; providing structured opportunities for graduate students to take active roles in the Center's research; and offering students, scholars, and citizens access to current theories, critiques and explications of the Internet." http://www.isc.umn.edu
- *Resource Center for Cyberculture Studies:* Description from its Web site: "The Resource Center for Cyberculture Studies is an online, not-for-profit organization whose purpose is to research, study, teach, support, and create diverse and dynamic elements of cyberculture. Collaborative in nature, RCCS seeks to establish and support ongoing conversations about the emerging field, to foster a community of students, scholars, teachers, explorers, and builders of cyberculture, and to showcase various models, works-in-progress, and on-line projects." http://www.com.washington.edu/rccs/

Another excellent source of information for doing digital media research are the many Web sites that attempt to study Web traffic, growth rates, user profiles, and so forth. These are good for keeping track of the latest usage statistics, although the methods for measuring Internet traffic are imprecise. One can compare and contrast the statistics from various sources to see whether they are consistent. There are many available these days, although not all the information is free of charge. One good place to start (because it is free and has reliable information) is Nielsen//Net Ratings (http://

www.nielsen-netratings.com). Right off the front page it tells you the top 25 Web properties, the top 10 banners, the top 10 advertisers, and global Internet usage.

How does Nielsen//NetRatings come up with its numbers? It explains on the site: "The reported Internet usage estimates are based on a sample of households that have access to the Internet and use the following platforms: Windows 95/98/NT, and MacOS 8 or higher. The Nielsen//NetRatings Internet universe is defined as all members (2 years of age or older) of U.S. households which currently have access to the Internet."

A handful of other sites on the Web provide statistics on Web usage. Public opinion survey sites, such as the Gallup Organization, (http://www.gallup.com), are helpful because researchers there often ask questions about Internet or Web use.

An important part of being a good researcher involves knowing what other researchers have done. Scholarship is not a solitary endeavor. Scholars do research and share their findings with others. In this way, the body of knowledge that is collected grows and benefits many people. Hence, even if you don't personally do research on digital media, you can certainly read what others have done by going to the library or the Web and searching for academic research about some aspect of digital media that you are interested in. If you do engage in research and feel that you've produced something worth presenting to the scholarly community, you may look into submitting it for presentation at a scholarly conference or for publication, although these venues are usually pursued by graduate students and faculty members. Ask faculty members or grad students about it. You may want to submit something collaboratively, or they can give you pointers about how to get your work and ideas disseminated to a wide audience.

Questions for Discussion and Comprehension

1. If you were going to embark on a research project relating to some aspect of digital media, what aspect of digital media would you be interested in and what might a question be that you would be interested in answering? *Before* you answer these questions, however, do a literature review of Internet-related research so that you get an idea of what has already been done.
2. Communications research often deals with effects of media on some segment of the population. What are some effects of using the Internet that might be investigated as part of a research project?
3. Go to the library and look for one scholarly article about the Internet, World Wide Web, online news groups, online community, or other aspect of digital media that catches your interest. Read it carefully, and then write a summary of the research, the findings, the conclusions, and so forth.

Endnotes

1. There has been so much written about Cultivation Theory, both by its adherents and its critics, that listing these sources would comprise a long bibliography in itself. Those interested in the theory should seek out Gerber's many articles as a starting point. A popular videotape called, "The Killing Screens: Media and the Culture of Violence" (Northampton, MA: Media Education Foundation, 1994) is also a useful resource to learn more about Gerbner and his work.

2. Guido H. Stempel III, Thomas Hargrove, and Joseph P. Bernt, "Relation of Growth of Use of the Internet to Changes in Media Use from 1995 to 1999," *Journalism and Mass Communication Quarterly*, Spring 2000, abstract.

3. Joseph R. Dominick, "Who Do You Think You Are? Personal Home Pages and Self-Presentation on the World Wide Web," *Journalism and Mass Communication Quarterly*, Winter 1999, abstract.

10

The Critical View

Fear and Fiction

For some, the pervasive growth of the digital media environment is a serious cause of concern, even alarm. Interestingly, these concerns are often articulated through fiction. Numerous movies dating back decades depict evil computers or evil people using technology to accomplish antisocial ends. The 1956 science-fiction thriller *Forbidden Planet*, for example, is about a series of mysterious murders on the planet Altair IV, murders that are later attributed to a homicidal technology in the planet's core. *Demon Seed* (1977) is about a computer that develops a criminal mind and becomes intellectually autonomous from its human creators.

Movies such as *2001: A Space Odyssey* (1968) and *War Games* (1983) suggest what happens when artificial intelligence goes out of control. *The Net* (1995) deals with identity theft and invasion of privacy issues. The list goes on and on. Although these are works of entertainment and not scholarship, they are an important part of our historical popular culture, which often reflects social fears and preoccupations through art. During wartime there are movies about war. *Guess Who's Coming to Dinner* (1967) was made at a time when interracial relationships were controversial. Likewise, the plethora of movies about computers and their negative impact on individuals and society possibly taps into a deep collective uneasiness about just where all this technology is taking society. Sometimes the world of fantasy and fiction is an ideal vehicle for contemplating the complexities of social phenomena and then advancing particularly salient issues for public discourse.

Perhaps one of the earliest critiques of telecommunications technology was a prescient work of short story fiction by E. M. Forster called, "The Machine Stops." In Forster's vision of a communicative space, humankind has burrowed itself under the surface of the earth and communicates via a global electronic network that carries both voice and video.

The protagonist of the story, an older woman named Vashti, lives in her little underground cell, physically isolated but filled with the trappings of high technology. She is said to know thousands of people, but her interactions with them occur only through an elaborate communications system that is empowered by something referred to simply as The Machine. The dramatic tension of the story arises when Vashti's mature son, Kuno, who lives "on the other side of the earth," begins to question the legitimacy of the dominant infrastructure through which most people on the planet communicate—an infrastructure remarkably similar in structure and process to the Internet (although the story was written in the early 1900s!).

"We say 'space is annihilated,'" Kuno tells his mother, "but we have annihilated not space but the sense thereof. We have lost a part of ourselves."In his sacrilegious critique of The Machine, Kuno concludes, "It has robbed us of the sense of space and the sense of touch, it has blurred every human relation."

Vashti initially dismisses Kuno's misgivings about the ubiquitous communications system. "I want you to come and see me," her son implores her, indicating that he has something important he needs to tell her. Looking at him through the "pale blue plate" (the interface through which they are communicating), she exclaims: "But I *can* see you! . . . What more do you want?" (1928, pp. 14–15).[1]

Throughout the story, mother and son debate the merits and demerits of an electronically mediated nonphysical space—the former heralding its ability to *enhance* community formation and human relations, the latter lambasting its deleterious effects on what Kuno considers genuine human interaction, which, apparently, he insists must include physicality and proximity. That the mother would be such a virulent proponent of the technology and her much younger son regarding the same with suspicion and cynicism suggests a world in which traditional perceptions about technological impact are turned upside-down.

Interestingly, the countervailing perspectives about community formation represented by Vashti and Kuno seem as relevant today as they were in the first decade of the twentieth century. Are emerging communication technologies community-building technologies?Or do they result in the segmentation and fragmentation of communities?Do they have, as communications scholar James Carey and others have called them, *centrifugal* (fragmenting) or *centripetal* (binding) effects on traditional communities? Or both? Or neither?Do they really link the world into a central nervous system, as McLuhan liked to describe it, resulting in a "global village," or do they merely give us the sense that space has been annihilated at the expense of "real" human interaction?

Social Criticism: Penetrating the Surface

Whenever new technologies appear in society, there is usually an accompanying chorus of voices that express concern and criticism of those technologies both in fiction and nonfiction. The period surrounding the emergence of movies and, later, television found editorials, opinions, academic articles, and other commentaries that expressed concern about the effect that these technologies might have on audiences. The current digital media environment is no exception.

The education of students interested in digital media would be incomplete without a section on social criticism. In the introduction to a book called *Muckraking Sociology* (1972), the editor, G. T. Marx, drew a relationship between social criticism and social change. He wrote that some kinds of social research can be incriminating, akin to the muckraking news stories of the late 1800s and early 1900s that documented and publicized social wrongs. "In pointing out the gap between values and actual practices,"

Marx wrote, "and in questioning established orthodoxies it serves as a vehicle for social criticism and, hopefully, social change."

Whether one agrees with any particular social critique or not, it is nevertheless important to be exposed to different points of view about social phenomena. In writings about the digital media environment in the popular press, one finds no lack of hype and enthusiasm. Digital media around the turn of the millennium represented the future and scientific and technological progress. And yet, the job of the scholar is not to propagate hype but to study, observe, research, think, and share. The scholar needs to see the "big picture," the flip side, the alternative scenarios and explanations in order to be truly "knowledgeable."

Social criticism, in whatever form it takes, helps us probe deeper into social phenomena and ask probing and sometimes disturbing questions. The social critic, a role that all scholars should embrace to some degree, is not satisfied with the conventional wisdom or the popular perception of things but with what lies beneath the surface. In the lingo of some social critics, they deconstruct dominant ideologies and expose the power relations these ideologies are meant to uphold. An ideology, as used here, is a body of ideas upon which a particular social, economic, and political system is based.

One could argue that a dominant ideology in the Digital Age is that digital media technologies are progressive, increase efficiency, and facilitate and democratize communication among members of civil society. They are, in the words of Ithiel de Sola Pool (1983), "technologies of freedom." A much different perspective is offered by Vincent Mosco in *The Pay-Per Society* (1989). Mosco views the emergence of a high-tech environment more suspiciously and cynically. In a world of electronic information systems, communication becomes commodified and privatized; rules and regulations are driven by an increasingly powerful private sector; technologies can be used for surveillance and control; and democracy and diversity suffer because content production falls into the hands of bottom-line capitalists. Mosco challenges the widely held notion that communication and information technologies liberate and empower communicators. In fact, he argues, these technologies can do just the opposite.

Whether one agrees with Pool or Mosco is less important than whether one is familiar with both of their arguments and can see the merits and demerits in each of their perspectives. Ultimately, one might lean toward one argument more than the other, but trying to get a full, balanced view of the problem at hand before arriving at a conclusion is the mark of a good scholar.

What follows are more current critiques of the digital media environment. They are provided here to give you a different view of the Digital Age from three people who are disturbed by some of the trends and developments they attribute to the high-tech trappings of the Internet, World Wide Web, e-mail, and so forth. These critics are not Luddites; in other words, they are not opposed to the use of technology. But they have legitimate concerns that should be aired to give readers something to think about and to advance public discourse about the Digital Age. Read these critiques with an open mind. If their concerns resonate with you, you might consider reading the whole book. If you disagree, you should formulate counterarguments in your own mind to their positions (as has been done here to some extent). Argument, even when impassioned, in the scholarly context is not usually meant to be malicious but rather educative to others and to oneself. Disagreeing with another person's point of view often forces one to refine and articulate one's own point of view. Critical analysis and argument are essential to good scholarship.

Criticism

Some good examples of critiques of the Digital Age include Clifford Stoll's *Silicon Snake Oil: Second Thoughts on the Information Superhighway* (1995), David Shenk's *Data Smog: Surviving the Information Glut* (1997), and Michael Noll's *Highway of Dreams: A Critical View Along the Information Superhighway* (1997). Unlike many other armchair critics of digital media, none of these authors is technophobic. Indeed, they are, by most accounts, technology experts and exceptionally well versed in the subject about which they write. They are the Vincent Moscos, Noam Chomskys, Todd Gitlins, and Ben Bagdikians of the *new* rather than traditional media environment. Like all good media critics, they attempt to reveal the subtext of social phenomena and force us to confront the underbelly of what might appear at first glance to be a good thing. These critics' views are representative of a larger body of criticism that asks us to take seriously the potential problems that accompany the digital media environment.

Stoll's *Silicon Snake Oil* is a contrarian's long essay about how digital media often rob, rather than contribute to, community formation among human beings. He laments the impending loss of face-to-face communication and waxes nostalgic for the days when things were less technology driven. He says he has "strong reservations" about online communities: "They isolate us from one another and cheapen the meaning of actual experience. They work against literacy and creativity. They undercut our schools and libraries" (1995, p. 3).

He continues:

> It'd be fun to write about the wonderful times I've had online and the terrific people I've met through my modem, but here I'm waving a flag, a yellow flag that says, "You're entering a non-existent world. Consider the consequences. It's an unreal universe, a soluble tissue of nothing." While the Internet beckons us brightly, seductively flashing an icon of knowledge-as-power, this nonplace lures us to surrender our time on earth. A poor substitute it is [where] important aspects of human interaction are relentlessly devalued. (pp. 3–4)

Stoll's criticism is a point well taken. It is hard to disagree with some of the more obvious claims he makes, such as that online communication is not the same as face-to-face communication. He suggests a number of bleak scenarios—a bookless library, a librarianless library, automated commerce, and so on—and laments the consequences to society. Although Stoll seems preoccupied with the negative aspects of technological progress rather than trying to put these changes in a broader social context, his concerns echo those who believe that digital technologies have a destructive undercurrent.

"The telephone eroded the art of writing letters," Stoll writes. "Television cut into neighborhood cinemas. MTV and superstars weakened amateur musicians and hometown bands. The car destroyed urban trolley systems; interstate highways devastated passenger rail service; and airliners wiped out passenger ships."

One could argue, as Stoll does, that new technologies often do make older technologies obsolete, but these new technologies often also lead to practical improvements that most people would consider socially valuable. For example, the telephone probably did erode the art of letter writing to some extent, but this same technology also facilitated more frequent, instantaneous, and reliable communication among people over long distances. Not everyone would regard letter writing as nostalgically as Stoll does, although certainly a hand-written letter now and then conveys a different "ethos" than a phone call or an e-mail message.

Nevertheless, the point Stoll seems to be making is an important one: Be careful what you replace for something "better." New is not necessarily better. Stoll's book serves as a reality check to overly optimistic declarations about the wonders of digital media. Too often, however, he tends toward the nostalgic yearnings of yesteryear. In describing the library of the future, for example, he sees "lots of books, a card catalog, a children's section with a story hour, a reading room with this morning's newspapers, plenty of magazines, a box of discarded paperback books (selling for a quarter each), a cork bulletin board stapled across with community announcements, a cheap photocopier, and a harried but smiling librarian." He also sees "a couple of library volunteers reshelving volumes" (pp. 175–178).

This seems more like the library of his childhood, perhaps a reminder of pleasanter or simpler times. Unfortunately, library space is not an unlimited resource. Card catalogs become increasingly unwieldy and inadequate as library holdings expand exponentially, as they invariably do. And what about those library volunteers? They could contribute to their community in other ways if just a portion of their work could be relieved through automated library systems and online access to *some* library information. Moreover, not everyone has physical access to a public library—because of a lack of time or because of where they live—so these happy outings to the library bothering the "harried but smiling" librarians are not realistic for a large number of people. They are better served if they can get information they need from their desktops.

Silicon Snake Oil, understandably, has received widely mixed reviews. Even librarians must wonder whether Stoll's defense of the traditional library was more of an unintentional backhanded compliment rather than a boost for librarianship. Librarians, after all, could benefit from the same technologies Stoll condemns to make their lives *less* harried; without them, their workdays would frequently verge on the unmanageable.

Stoll is no Luddite. He is both a scientist by training and a knowledgeable computer-user by practice. Why does he have "second thoughts" about the Information Superhighway? Like so many people whose bodies seem tethered to a computer for long periods of time, he probably understands how easy it is to get overly engaged with technology to the detriment of one's real-life, face-to-face social interactions. People certainly do need to beware of the isolating tendencies that Stoll fears, but striking a balance in life applies to many things, not just computer use. Few would disagree that computer addiction is a personal and social detriment. Although somewhat extreme in its examples, his book serves as a good reminder that communication involves human relationships, not relationships with machinery.

Shenk's *Data Smog* has much in common with *Silicon Snake Oil*, except Shenk's critique is perhaps more relevant to online news and information than digital media in general. Shenk quotes media critic Neil Postman at the very beginning of the book:"We have transformed information into a form of garbage." He then quotes Bill Clinton, although the remark seems oddly out of context, given the former president's strong support for a national information infrastructure since the earliest days of his administration. Says Clinton: "In the information age, there can be too much exposure and too much information and too much sort of quasi-information.... There's a danger that too much stuff cramming in on people's minds is just as bad for them as too little, in terms of the ability to understand, to comprehend."

Shenk writes about a problem that many people have complained about in their day-to-day lives. They are feeling overwhelmed by the amount of information around them. Pretty soon, some argue, one doesn't know what's important to know and retain and what's not:

> At a certain level of input, the law of diminishing returns takes effect; the glut of information no longer adds to our quality of life, but instead begins to cultivate stress, confusion, and even our ignorance. Information overload threatens our ability to educate ourselves, and leaves us more vulnerable as consumers and less cohesive as a society. For most of us, it actually diminishes our control over our own lives, while those already in power find their positions considerably strengthened. (1997, p. 15)

Data smog is a term that frames the wider and greater accessibility to information as a pejorative. But it is also matter of semantics. His "data smog" may be another's information liberation, enlightenment, or refreshing abundance. Shenk tries to convince the reader that too much information creates too much confusion for society. This may be an outcome for some people, but the solution probably lies in helping people to manage and organize their information rather than complain that there is too much of it. It is like an obese person complaining that the source of his weight problem is the overabundance of food in the world from which to choose. Clearly, one doesn't have to eat everything, nor does one have to avail himself or herself of all the information that is available in the world. We learn to pick and choose. Not to do so would likely lead to the kind of overwhelming stress that Shenk describes well in his book.

Michael Noll's *Highway of Dreams* looks at all the digital media hype and is refreshingly unimpressed. In practically the same breath, he declares the Information Superhighway a bunch of "hype and fantasy" and then announces that, in many ways, the superhighway is already here. His analysis is centrist: The utopian interpretation of the superhighway, he believes, is untenable, but the current communications environment—the switched public telephone network, industrial private networks, public and private packet-switched data networks, CATV cable to the home, and satellite communications—*already* comprise an information superhighway.

Noll, a former Bell Labs researcher and current professor at the Annenberg School of Communications at the University of Southern California, puts the hype surrounding the Information Superhighway into techno-historical context. In his experience as a technology researcher, policy analyst, and educator, he has seen great ideas for new technologies come and go, regardless of the hype surrounding them, for obvious reasons: cost, complexity, lack of consumer need, undeliverable promises and so forth. After reciting a high-tech scenario for the reader, comprising all the familiar promises of digital media grandstanders, Noll reenters with this sobering reality check:"It all sounds great, but if you are as old as I, much of it sounds all too familiar. You wonder whether people will really want all these services, whether they will [be able to] afford them, whether industry will really be able to deliver them or afford the required investment, and whether governmental policies will allow them" (1997, p. 13).

Noll casually wanders down the road of technological history and attempts to use history to show *not* that what's coming is dangerous but that what's coming is *not* coming, at least on the grand scale that many people have been led to believe. At the end of his book, he realizes he may not have won many converts: "Although I have attempted to convince you otherwise, perhaps you still are convinced that the communication superhighway is coming and will be the dawn of a new, revolutionary era and society. I hope that our journey has educated, informed, and convinced you that the superhighway is mostly superhype.... [A]void the superhighway" (p. 193).

Although Noll's arguments are informative and thoughtful, his book is probably more successful at reeling in extreme points of view rather than convincing readers that the Information Superhighway is just a highway of dreams. For one thing, some of what he seems to dismiss as utopian fantasy is already here. Early in his book, he describes some future scenarios: "Hundreds of channels of television programming from around the world will be sent to our homes from communication satellites located in geosynchronous orbits 22,300 miles above the earth's equator. A small antenna at our home will be electronically steerable to aim itself to receive the signals from the satellite" (1997, p. 6). The growing popularity of direct broadcast satellites (DBS), which, granted, are not yet the norm in U.S. households is less of a fantasy than many of his other examples. It may not be a viable competitor to cable television right now, but it is possible it will catch on in the future as satellite communications becomes more commercialized for individual consumers. His conservative view of low earth orbit satellites (LEOs) is proving to be more accurate than not, however. The industry has had to scale back its vision considerably since the mid-1990s.

Noll's book is reminiscent of one written years ago, Steven P. Schnaars's *Megamistakes: Forecasting and the Myth of Rapid Technological Change* (1989). Schnaars raises some of the same arguments that Noll raises, and uses many of the same examples. He asserts that, in general, change occurs gradually and conservatively, not dramatically, and that years from now it will still be news if a duke marries a dustman's daughter, just as it has been in centuries past. History reveals, he says, that grandiose high-tech future scenarios tend to be wrong.

Noll and Schnaars may be right. But they also may be wrong. Many technologies that are commonplace today may have seemed highly unlikely to penetrate the mass consumer market at one time—the personal computer being just one example, but also the fax machine, cellular phones, personal paging devices, automated bank machines, laser code scanners, even microwave ovens. The conservative outlook is not necessarily the most compelling; it is simply safe. Technological change in society occurs regularly, but it often occurs with fits and starts, sometimes requiring a series of graduated trials over time before successfully penetrating the mass consumer market. Pinpointing when that penetration occurs, obviously, is not easy because so many interrelated factors affect it. The conservative outlook will be correct a good amount of the time simply because there are forces in society that benefit from maintaining the status quo. When the status quo is challenged by technology (e.g., the printing press in fifteenth-century Europe, the telegraph and wireless communications in the eighteenth and nineteenth centuries, and computer networking technologies in the twentieth and twenty first centuries), significant social changes do occur.

Schnaars dismisses a number of technologies whose ascendancy was probably only in remission, not dead, when he was studying them. Distance education is one example. At one time, distance education could have posed a major economic and philosophical threat to established institutions of higher education. It would upset the status quo. Indeed, colleges and universities themselves were generally dismissive of distance education. Now that distance education has been gradually appropriated by a more market-oriented educational system, it is seen as a viable alternative revenue stream with a potentially lucrative target audience. This time around, it may actually hit the ground flying.

The point is that Schnaars and Noll don't see the digital media environment with all its online interactive and conversational potential to be viable at this time, if ever, for a variety of reasons. Stoll and Shenk acknowledge that the change is here but they don't like what they see coming. All of these perspectives lend context and perspective to the hype that too often surrounds digital media technologies.

Noll's book, like Stoll's and Shenk's, is valuable because it advocates temperance and reason when studying the digital media environment and points the reader to an Information Superhighway that is not a fantasy of the future but a reality of the present. He forces the reader to think twice before jumping on the digital media bandwagon and implores the reader to study history for clues about the future of new technology.

These are only three books that take a critical eye to the digital media environment. There are others. It is important to read what the critics have to say because they help people think critically about social phenomena. Good scholars are critical thinkers.

Framework for Critical Inquiry

One of the best ways of becoming a critical thinker is by questioning and challenging established notions and the so-called conventional wisdom. If people throughout history always accepted the answer, "These are the way things are and will always be," social change would never occur. Instead, some people asked *why*. Who are the beneficiaries of the status quo (i.e., leaving things the way they are) and who are the losers?

The following questions provide a critical framework for analyzing social conditions, phenomena, and ideologies. They are meant as a way to think about developments in the Digital Age more complexly and interestingly, in a multidimensional sort of way. Because technological change does not occur in a vacuum, it is important to look at many different ways that technological change affects society. Part of that involves questioning information, ideas, and ideology. Here are some ways to start:

- What/Who is the *source* of the information? Is the source affiliated with a corporation, an educational institution, a nonprofit organization, a special-interest group, a government entity, or some other interest? If so, what motivations could

the source have for presenting information in a particular way (i.e., emphasizing some points and ignoring others)?
- Are there other perspectives out there that should be considered? What are they? Do you find opposing or alternative arguments and explanations credible?
- Is there a dominant ideology being reinforced? What is that dominant ideology? Do you agree with it? Are there hidden agendas that need to be exposed?
- Who are the "winners" and "losers" in society if a particular future scenario is realized? Are existing disenfranchised or underserved groups helped or hindered by pending social and technological developments?
- How can things be improved? What safeguards and protections need to be implemented for positive social change to occur?
- What do people need to do to improve society? What needs to happen? What needs to stop or be prevented? What needs to be communicated?
- Who are some of the social critics who have relevant, important things to say on the matter? What are they saying? What are their prognostications?

Adopting a healthy skepticism and critical eye toward the digital media does not make one a Luddite or antiprogressive. Instead, it makes one broad-minded, sensitive to the needs of those who do not have social power, and able to advocate for and contribute to appropriate social change that benefits many rather than only a few. As social activists and others rise up around the world against what they perceive to be the abuses of the global corporate culture, proponents of technological progress in the Digital Age need to be mindful of how technology can be both helpful (e.g., facilitating community building) and hurtful (e.g., further alienating the poor and disadvantaged). A scholar should not ignore these dichotomies but should expose them and challenge them head-on, helping to affect social change in productive and compassionate ways.

Questions for Discussion and Comprehension

1. The authors discussed in this chapter pose serious concerns and criticisms about digital media that they would like readers to consider. Do you agree or disagree with them? What are other concerns that you think people should be thinking about?
2. What has your experience been with digital media thus far? Do you think, for example, that the Internet has been a hindrance in your life (e.g., takes up too much of your time)? Or has it benefited you (e.g., helped you with research or kept you in touch with friends)? Explain.
3. See if you can get one of the books discussed here (either from the library, from a teacher, or from a bookstore) and read it in full. Write a book report that analyzes the author's perspective and discuss it with your classmates. Or find another book that is critical of the digital media environment and do the same thing.

Endnote

1. Although the collection of stories from which this is taken was published in 1928 (see bibliography), this specific story ("The Machine Stops") was written on a date "prior to 1914," according to the author. Emphasis added.

11

The Future

Rethinking the Digital Future

Many Americans look to the future with uncertainty and a certain amount of fear. The nation's security and economy are closely intertwined, and both face challenges in the years ahead.

Where digital media and the economy are concerned, one thing is certain: The hysteria surrounding the digital revolution of the 1990s has subsided. Top executives at even the most economically viable media conglomerates are demanding accountability from their new media divisions, once thought to be the coddled upstarts that would grow up and take their parent companies into the twent first century. Now some of these digital media arms seem in danger of amputation.

One could argue that the portent to today's retrenchment can be traced back to the "Mother of All Media Web Sites"—namely, Time Warner's gargantuan Pathfinder project. Born in 1994, Pathfinder was supposed to be the online showcase for (and digital portal to) Time Warner's many brand-name publications such as *Time*, *Fortune*, *Money*, *Entertainment Weekly*, and *People* magazines. According to a former staff member who spoke to the *New York Times*, the company was spending $15 million a year to develop the site.[1] After years of hemorrhaging money into this massive online project, the media conglomerate was still not recouping anything from its investment and became increasingly skeptical that it ever would.

Without nearly as much fanfare as when the Web site launched, Time Warner began breaking up Pathfinder in late 1998, moving some of its content to an upstart digital media company called America Online, at that time mainly known as burgeoning Internet service provider. (This was two years before the announcement of a Time Warner/AOL merger.) Finally, in early 1999, Time Warner decided to shut down the beleaguered Pathfinder altogether. The individual magazines would have their own Web sites, but the portal was history.

Lessons and Insights

The Pathfinder experiment provided a number of valuable insights for traditional media organizations branching into online services. First, the dominant players in the traditional world of mass media do not necessarily reign in cyberspace. Relatedly, the "brand names" of traditional mass media with big audiences in the real world are not guaranteed a comparable following online.

Second, no matter how wealthy a company is (Time Warner had deep pockets as one the world's largest media conglomerates), there is a limit to the amount of resources it will commit to digital media development. Although the cheerleaders of digital media in a company may promise eventual returns on heavy upfront and continuing investment, media executives will not hesitate to turn off the gravy chain when those returns do not materialize. Talk is cheap; accelerated digital media projects are expensive.

Third, the digital media audience may not be attracted to a simple repurposing of content that appeals to a more traditional, off-line audience. The concept of *relative advantage* (i.e., providing the customer with sufficient additional features and benefits above and beyond the traditional product or service) is difficult to apply in concrete terms. Some media organizations thought their brand names were powerful enough to carry them ahead of their online competitors in terms of audience popularity without paying much attention to relative advantage. They were wrong.

Fourth, the commercial failure of digital media arms of traditional media organizations has largely been due to the inability to demonstrate a solid business model. Advertising-driven models online have turned out to be a shaky business proposition at best. It has long been known that banner ads are ineffective, but what else is there to take their places? What are the viable alternative revenue streams, if not advertising? The subscription model has been tried and abandoned, except in rare cases (e.g., pornography and specialized business news) where there is some evidence of success. Micropayments (i.e., paying the customer small amounts to view ads) has all but disappeared after a relatively brief experimental stage. E-commerce (i.e., shopping and purchasing online) may have potential but is not proven. In short, online profits are hard to come by, and fresh ideas are few and far between. Companies that grew their digital media divisions too quickly, expecting financial rewards that never came, need to "ungrow" to stop continued losses.

Fifth, newcomers to media such as Yahoo!, America Online, MSN, and others may be more in tune with the Net Generation (if there is such a thing anymore) than the old money of the traditional media world. Marriages between old and new, however, might comingle the experience and maturity of traditional media with the innovation and vivaciousness of digital media to form a third entity that integrates the best of both worlds. One of the problems with many floundering dot-coms is that they grew too large too quickly and were grossly overvalued, which resulted in plummeting stock prices.

The Current Situation

These lessons' insights are useful for illuminating more recent developments in the digital media world that are spreading virus-like from one major media organization to another in a fairly short period of time. Clearly, top executives at some of the most powerful media conglomerates have become impatient and unhappy with their digital media spin-offs. Rupert Murdoch's News Corporation announced in the first week of

January 2001 that the company planned to lay off several hundred workers in the coming weeks at its New York and Los Angeles offices of FoxNews.com, FoxSports.com, and Fox.com sites on the Web. The action was taken to save the company tens of millions of dollars each year. News Corporation actually began dismissing digital media staff in October 2000, not long after Murdoch himself announced at an annual meeting that he had lost confidence in an advertising-based Internet business model.[2]

The Walt Disney Co.'s chairman, Michael Eisner, announced on January 29, 2001, that Disney was shutting down its Web portal, Go.com, laying off 400 employees, and would take a $790 million write-off. (Disney also controls ABCnews.com and ESPN.com but said it would not take any action against those sites. Disney also owns the ABC television network.) A little more than a year earlier, Eisner predicted that the Go.com portal would "be the next Nirvana for us" and said the company planned on spending a billion dollars on promoting the site.[3] Nirvana never arrived. The Walt Disney Internet Group has never been profitable and was expected to post losses when its most recent quarter-end report is announced.

Disney is not the only media company that has attempted to create a popular—and profitable—Web portal. NBC tried to do the same with its NBCi, a project that seems to be on a slow death march. In August 2000, the company cut 170 jobs. A second round of layoffs was announced in January 2001—this time 150 jobs, or 30 percent of its workforce.[4] Having laid off a large portion of its workforce and shut down most of its operations, NBCi, keeping in line with an industry trend, is no longer the dynamic portal the NBC network once hoped it would be.

Perhaps most alarming for sheer numbers was AOL Time Warner's announcement in January 2001 that it would lay off of 2,400 employees and sell its Warner Bros. retail stores or close them if a buyer cannot be found. If the stores are closed, 3,800 employees could be out of a job. The layoffs are the result of the recent merger between AOL and Time Warner.

Like a growing number of their dot-com cousins in the nonmedia business, many traditional media organizations have seen stocks in their digital media spin-offs plummet in recent months. The problem is not new. In mid-August 2000, the *New York Times* reported that Wall Street has "all but written off the broadcast companies as Internet players." It said that the publicly traded shares of NBC Internet and the Walt Disney Internet Group were down more than two-thirds and that experience has been "a humbling comedown for the networks and their owners, who are more accustomed to being mighty media monoliths than niche players scrounging for audiences."[5]

Newspapers have also rethought their new media ventures. In mid-October 2000, the New York Times Co. changed its mind about taking its Internet business division public. It decided to withdraw its application to the Securities and Exchange Commission on the same day that the Tribune Co. announced cutbacks at its online media division. Although business seemed to be improving for New York Times Digital, it was projected to lose $62 to $65 million dollars in 2000.[6] On January 7, 2001, the company said it would lay off 70 employees. The cuts would affect a number of different Web sites, including nytimes.com, nytoday.com, and boston.com. Previously, the newspaper publisher Knight-Ridder said it would lay off about the same amount of people from its digital media division.[7]

These are but a smattering of examples of the digital retrenchment currently occurring in some of the nation's largest media organizations. The list is suggestive, not exhaustive, and begs the question: What does it all mean? Are traditional media turning their backs on digital expansion and retreating to the safety of convention? Has the "digital experiment" been branded a dismal failure and relegated to the end of the priority list of things-to-do in large media corporations?

The answer is both yes and no.

Regrouping but Not Departing

The current retrenchment suggests that media companies are rethinking (1) the rate of growth of their digital media enterprises and (2) the specific types of content that they should try to market to an Internet audience.

One does not have to look too far to find valuable lessons about the dangers of rapid growth and limp profits. The business news is replete with articles about how once-overvalued dot-coms have scaled back their vision and expansion in response to the harsh economic realities facing them on the pages of their quarterly ledgers. Kozmo, drugstore.com, Avenue A, and dozens of others have cut staff and restructured, which many financial analysts consider to be economically prudent. Others, such as Hardware.com, MyLackey.com, Onecast Media, EZBid.com, Mercata, Bazillion, and Go.com, have simply closed shop. What happened to Petstore.com, which hoped to be the Amazon of the pet supply market, is not atypical. It went public on NASDAQ, bankrolled a national advertising blitz, benefited from the dot-com euphoria for a short while, and then saw its stock plummet from 14 dollars to 22 cents. The stock dropped 98 percent in about 10 months. The *Industry Standard*, a publication that reports on the health of the Internet economy, has been keeping track of layoffs at dot-coms since December 1999. By January 19, 2001, it had tracked 43,192 layoffs.[8] A week or so later, it would have to add the 400 layoffs at Go.com (mentioned previously) and the 1,300 jobs Amazon.com announced at the end of January that it was cutting. (Ironically, the *Industry Standard* itself has had to cut jobs on its own staff.)

Media organizations, however, are in a different position than dot-com start-ups such as Petstore.com or even Amazon, which started out as exclusively online enterprises and had nothing to fall back on when the going got rough. Media organizations are not in the same kind of all-or-nothing situation that dot-com start-ups are in; if the digital media division of a media organization is unprofitable, it can be subsidized by the parent company, which oversees a diverse array of subsidiary companies and can shift money from one place to the other as it sees fit. Hence, whereas lack of profits could send a dot-com start-up out of business, new media divisions of traditional media organizations may survive, but not necessarily as once envisioned.

Despite signs that large media companies are backpedaling on their commitment to their digital media enterprises, it is highly unlikely that these companies are going to abandon their plans for expansion into digital media altogether in the long term. These companies are clearly in a serious cost-cutting mode at present, which is a dramatic shift from the days of almost blind-faith investment in their new media projects. All the large media companies seem to be trimming their digital media plans to some extent, but they are *not* eliminating them. Disney realized it was not going to be competitive in the Web portal business (just as Time Warner had concluded a few years earlier), and so it killed off that very expensive component of its digital media program, but it will continue to operate Disney.com, ABC.com, ABCNews.com, and NFL.com, and may even sprout other ventures later on. The other companies mentioned earlier are doing the same. There is a strategy of "selective retrenchment" currently occurring, which in the long term is probably good for digital media, although in the short term clearly distressing for those laid off from their jobs and probably even for their employers. No doubt many employees feel they have been sacrificed to Wall Street, cut off the payrolls along with a number of their colleagues to increase profitability by reducing costs paid in salaries. It is not unusual for stocks to increase in value after layoff announcements.

What you hear over and over again from companies that have not fared well with digital media is this: They were too ambitious in their original business plans. Advertising revenues were sagging. They hired too many people. Downsizing is a way of streamlining costs. Market conditions have changed.

In the long term, it is unlikely that the digital media enterprises of traditional news organizations are going to suffer irreparably from the current retrenchment. In fact, the more likely scenario is that just the opposite is true. Like a wild animal with its foot caught in a trap, sometimes it is more beneficial to chew off part of the leg than to let the whole body die. Digital media enterprises were always a risky business, for both the companies that invested in them and the people who chose to work for them. At first, it seemed like a gold rush, but when the gold is nowhere to be found, what is a business to do? The answer seems to be: Step back, reassess, and move forward cautiously. Some companies seem to be in a holding pattern or are going on with business as usual. (USA Networks' CitySearch and the Washington Post Co. fit into this category.)[9] Others are rethinking their content, marketing, and revenue-generating strategies. Still others have cut their losses and are focusing on their traditional strengths.

But the Internet isn't going anywhere. The vast potential and opportunity for traditional media organizations to exploit this technology still exists; the problem that remains to be solved is how to do it effectively—and for businesses (and even news organizations are businesses, after all), that means *profitably*. Under current conditions, selective retrenchment and the scaling back of online ventures make sense. Many digital media divisions had promised to show profitability by now, and that is not occurring. Nor does it appear that a sudden critical reversal toward profitability is on the horizon.

Allan Sloan, writing for the *Washington Post*, believes that many people have been living in a Dreamland for the past several years and that the current conditions discussed in this article are really just a return back to normal. He writes, "Uncertainty is in the stock market, layoffs are in the *zeitgeist*. Welcome to the real world. . . . You can't have all boom all the time. Life doesn't work that way. But don't go from irrational exuberance to irrational depression without at least a brief stay in the middle. The world isn't coming to an end."[10]

Sloan makes a lot of sense. Although it is difficult to ignore the considerable news coverage of ailing dot-coms and digital media experiments, one should not overreact to this news by assuming that the digital media industry is dead, or even plodding closer in that direction. What financial analysts often euphemistically call a "correction" when stocks plummet on Wall Street is actually a good description of what is happening with digital media. Traditional media companies are facing the fact that may have been wrong when starting on this journey a few years ago, and now they are making corrections.

Along with their personnel cuts, AOL/Time Warner and other media companies are integrating their digital media employees more with their traditional media division in an effort to eliminate redundancies and create efficiencies that did not exist when walls between the traditional and digital media operations were virtually impenetrable. No doubt that this is based chiefly on economics, but it could have important philosophical implications as well. Too often, the traditional or "old" media division and the digital or "new" media division have been kept apart, as if separating entities that were unlikely to get along. This seemed to create counterproductive identities in the media organization—one side identifying itself with the traditional division, and the other side with the digital division. Why should these parts be separate? In many media companies where layoffs have occurred, a more integrated personnel is the goal. This creates efficiencies but also conjoined identities and loyalties, and perhaps should have been the structural model to begin with.

Moreover, whether good or bad, the reality is that for many companies, stock performance, quarterly earnings, and profit projections are integrally tied together. With the stock market as volatile as it is at present and the possibility that the country is headed toward a recession, companies that are not profitable often have to make some brutal decisions about staffing, where much of a company's cost is centered. After circumventing the issue for years, Amazon has finally projected profitability by the fourth quarter of 2001. The company did not believe it could meet that goal with its pre-layoff workforce. Joel Naroff, an economic consultant quoted in the *Seattle Post-Intelligencer*, maintained that cutbacks, although personally difficult, are necessary for companies to remain competitive. He said, "Ultimately, the adjustments that the economy is making is going to set us up for the next strong period of growth."[11]

Media organizations such as News Corporation, AOL/Time Warner, Disney, the New York Times Co., the Tribune Co., and others are making the adjustments that they believe will allow them to regroup after an extended period of unfulfilled financial expectations from their digital media projects. During this restructuring and hiatus from aggressive growth in digital media, they will have a lot to think about—sort of a "Where do we go from here and how and why?" kind of period of systematic reflection. So many mistakes, miscalculations, and missteps litter the past few years for traditional media companies that tried to launch successful digital operations. Ironically, the start of the new millennium, which often was heralded as the bearer of great change and the digital revolution, really seems to be downtime for companies to think seriously about structure, content, and strategy before launching the next round of digital business experiments.

Only the most shortsighted of these companies would give up now, with an ever-growing audience on the Internet spending record amounts of money and surfing for content that traditional media organizations are in the business of creating. Nevertheless, now is not the time for reckless expansion either. Media companies did what they felt they had to do, and now they have to figure out what to do next to make things better. This is neither the best of times nor the worst of times for the traditional media business. The decisions that get made from here on, however, might tip the scales in one direction or the other.

Conclusion

This book has provided a fairly comprehensive overview of the digital media environment and its impact on traditional media and on society in general. It has been a narrative with ups and downs. Digital media have been called the "fifth news medium" by those who think it will rival the traditional news media of newspapers, magazines, radio, and television. Indeed, in less than a decade, the digital media themselves have become top news stories, heralding the promises and perils of an emerging and unpredictable communications environment.

In some ways, as I hope this book has shown, the digital media are not the bewildering and revolutionary phenomena that they are sometimes made out to be. Rather, they provide yet another way—to borrow Lasswell's constructs—that society has created to survey the environment, correlate the components of society in responding to the environment, and transmit social heritage. Or, as Schramm simplified, to be watcher, forum, teacher, and (he added one more) *entertainer*.[12] As the town square and village commons once set the stage for exchanging news and information among those who gathered there—in physical space—the traditional and digital media have set the stage for those who gather in cyberspace. Both of these "spaces" are important to the formation of a public—a public sphere and a community—and ultimately to the maintenance of an informed and empowered citizenry in a state of democracy.

On the other hand, however, the digital media are a bewildering and revolutionary phenomena. Cyberspace is an unprecedented communication and information environment. No one is quite sure how to deal with it, how to predict its impact on society, or even how to talk about it. As a result, as with many things new—particularly technology—the digital media have come under intense scrutiny, been regarded by some with suspicion, and, worse, have simply been dismissed altogether. This last approach, I believe, is misguided and detrimental. Digital media will continue to transform the communications landscape for many years into the future. In these times of change and uncertainty, it is likely that people will want much more, not less, news and information from digital media.

Questions for Discussion and Comprehension

1. Should traditional media organizations that have lost money from their digital media experiments abandon their digital media projects, keep advancing with their plans, or do something else? Choose an organization that fits this description, suggest a course of action, and explain why.

2. How would you describe what *cyberspace* is to someone who is unfamiliar with digital

media? How does cyberspace resemble the real world, and how is it different?

3. If you had to speculate about the future of digital media as an economic, social and political force in society, what would you have to say? How will media and society be changed in the Digital Age?

Endnotes

1. Alex Kuczynski, "Time Warner to Shut Down Its Pathfinder Site on the Web," *New York Times*, April 27, 1999, p. C-1. Time Warner itself has always been reluctant to officially discuss the financial costs of Pathfinder.

2. Jayson Blair, "News Corp. Plans Hundreds of Layoffs at Web Sites," *New York Times*, January 5, 2001, p. B-1.

3. http://www.msnbc.com/news/523120.asp

4. Saul Hansell, "More Jobs Cut at NBCi," *New York Times*, January 19, 2001, p. C-3.

5. Saul Hansell, "The Medium Gets the Message: TV's Monoliths Have Learned the Web Is a Fragmented World," *New York Times*, August 14, 2000, p. C-1.

6. Christopher Stern, "New York Times Pulls Plug on Internet Spinoff," *Washington Post*, October 13, 2000, p. E-03.

7. Felicity Barringer, "New York Times Company to Cut Jobs at Internet Unit," *New York Times*, January 7, 2001, section 1, page 23.

8. Mark Jurkowitz, "Online News Outlets Catch Their Breath: Cyber Slowdown Prompts Re-Thinking," *Boston Globe*, January 19, 2001, p. D-1.

9. David Lieberman, "Media Firms Trim Net Operations Amid Slowdown," *USA TODAY*, January 30, 2001, p. 3-B.

10. Allan Sloan, "Return from Dreamland Doesn't Mean We're in a Nightmare," *Washington Post*, Janauary 30, 2001, p. E-3.

11. Seattle Post-Intelligencer Staff and News Services, "Some Fear Layoffs Will Drag Down Economy," *Seattle Post-Intelligencer*, February 1, 2001, Business Section.

12. For early essays espousing on the structure and function of mass media, see Schramm's (1960) *Mass Communications: A Book of Readings.*

Bibliography

Bandura, Albert. (1977). *Social Learning Theory.* Englewood Cliffs, NJ: Prentice-Hall.

Bell, Daniel. (1976). *The Coming of Post-Industrial Society: A Venture in Social Forecasting.* New York: Basic Books.

Benedikt, Michael. (1991). *Cyberspace: First Steps.* Cambridge, MA: MIT Press.

Blumer, Herbert, and Philip M. Hauser. (1933). *Crime, Movies and Deliquency.* New York: Macmillan.

Bronson, Po. (March 1996). "Does He Really Think Scarcity Is a Minor Obstacle on the Road to Techno-Utopia?" *Wired,* pp. 124–126, 186–188, 192–195.

Campbell-Kelly, Martin. (1996). *Computer: A History of the Information Machine.* New York: Basic Books.

Carey, John, and Mitchell L. Moss. "The Diffusion of New Telecommunication Technologies." *Telecommunications Policy, 9,* 145–158.

Ceruzzi, Paul E. (1988). *A History of Modern Computing.* Cambridge, MA: MIT Press.

Compaine, Benjamin. (1984). *Understanding New Media.* Cambridge, MA: Ballinger.

Compaine, Benjamin. (1988). *Issues in New Information Technology.* Norwood, NJ: Ablex.

Curtin, Philip D. (1986). *Cross-Cultural Trade in World History.* New York: Cambridge University Press.

Fidler, Roger. (1997). *Mediamorphosis: Understanding New Media.* Thousand Oaks, CA: Pine Forge Press.

Folkerts, Jean, and Dwight L. Teeter. (1989) *Voices of a Nation: A History of the Media in the United States.* New York: Macmillan.

Forester, Tom. (1987). *High-Tech Society: The Story of the Information Technology Revolution.* Cambridge, MA: MIT Press.

Forster, E. M. (1928). "The Machine Stops." In *The Eternal Moment and Other Short Stories.* New York: Harcourt, Brace & Co.

Frederick, Howard. (1993). *Global Communications and International Relations.* Belmont, CA: Wadsworth.

Gilder, George. (1989). *Microcosm: The Quantum Revolution in Economics and Technology.* New York: Simon and Schuster.

Gilder, George. (1992). *Life After Television.* New York: W. W. Norton.

Hall, John A. (Ed.). (1995). *Civil Society: Theory, History, Comparison.* Cambridge, England: Polity Press.

Harris, Michael H. (1984). *History of Libraries in the Western World.* Metuchen, NJ: Scarecrow Press.

Kidwell, Peggy Aldrich. (1994). *Landmarks in Digital Computing: A Smithsonian Pictorial History.* Washington, DC: Smithsonian Institution Press.

Malamud, Carl. (1992). *Exploring the Internet: A Technical Travelogue.* Englewood Cliffs, NJ: PTR Prentice-Hall.

Marx, G. T. (Ed.). (1972). *Muckraking Sociology.* News Brunswick, NJ: Transaction Books.

Mirabito, Michael M. A. (1994). *The New Communications Technologies* (2nd ed.). Boston: Focal Press.

Moran, Barbara B. (1984). *Academic Libraries: The Changing Knowledge Centers of Colleges and Universities.* Washington, DC: Association for the Study of Higher Education.

Mosco, Vincent. (1989). *The Pay-Per Society: Computers and Communication in the Information Age.* Norwood, NJ: Ablex.

Naisbitt, John. (1982). *Megatrends: Ten New Directions Transforming Our Lives.* New York: Warner Books.

"National Information Infrastructure: Agenda for Action" (September 1993). Washington, DC: U.S. Government Policy Report.

Negroponte, Nicholas. (1995). *Being Digital.* New York: Knopf.

Noam, Eli M. (1995). "Electronics and the Dim Future of the University." *Science, 270,* 247–249.

Noll, A. Michael. (1997). *Highway of Dreams: A Critical View Along the Information Superhighway.* Mahwah, NJ: Lawrence Erlbaum.

Pavlik, John. (1996). *New Media Technology: Cultural and Commercial Perspectives.* Boston: Allyn and Bacon.

Pool, Ithiel de Sola. (1983). *Technologies of Freedom.* Cambridge, MA: Belknap Press.

Porat, Marc U., and Michael R. Rubin. (1977). *The Information Economy.* Washington, DC: Government Printing Office.

Rheingold, Howard. (1993). *The Virtual Community: Homesteading on the Electronic Frontier.* Reading, MA: Addison-Wesley.

Schnaars, Steven P. (1989). *Megamistakes: Forecasting and the Myth of Rapid Technological Change.* New York: The Free Press.

Schramm, Wilbur. (1960). *Mass Communications: A Book of Readings.* Urbana: University of Illinois Press.

Shenk, David. (1997). *Data Smog: Surviving the Information Glut.* San Francisco: HarperEdge.

Stoll, Clifford. (1995). *Silicon Snake Oil: Second Thoughts on the Information Highway.* New York: Doubleday.

Tarde, Gabriel. (1969[1901]). "The Public and the Crowd." In T. Clarke (Ed.), *Gabriel Tarde: On Communication and Social Influence.* Chicago: University of Chicago Press.

U.S. Department of Commerce, Economics and Statistic Administration, Office of Policy Development. (June 2000), *The Digital Economy 2000.* Washington, DC: Author.

Index

N

O

P

R

S

T

U

V

W

CHILDREN'S WELFARE AND THE LAW

The Limits of Legal Intervention